Methods in

CYCLIC NUCLEOTIDE RESEARCH

METHODS IN MOLECULAR BIOLOGY

Edited by

ALLEN I. LASKIN
ESSO Research and Engineering Company
Linden, New Jersey

JEROLD A. LAST
National Academy of Sciences
Washington, D.C.

VOLUME 1: Protein Biosynthesis in Bacterial Systems, *edited by Jerold A. Last and Allen I. Laskin*

VOLUME 2: Protein Biosynthesis in Nonbacterial Systems, *edited by Jerold A. Last and Allen I. Laskin*

VOLUME 3: Methods in Cyclic Nucleotide Research, *edited by Mark Chasin*

VOLUMES IN PREPARATION

VOLUME 4: Nucleic Acid Biosynthesis, *edited by Allen I. Laskin and Jerold A. Last*

VOLUME 5: Subcellular Particles, Structures, and Organelles, *edited by Allen I. Laskin and Jerold A. Last*

VOLUME 6: The Immune Response at the Cellular Level, *edited by Theodore P. Zacharia*

VOLUME 7: Peptide Hormones: Structure and Function, *edited by Bryan Brewer*

Methods in
CYCLIC NUCLEOTIDE RESEARCH

Edited by MARK CHASIN
*The Squibb Institute for
Medical Research
Princeton, New Jersey*

MARCEL DEKKER, INC. New York 1972

MARCEL DEKKER, INC.

95 Madison Avenue, New York, New York 10016

LIBRARY OF CONGRESS CATALOG CARD NUMBER: 72-89528

ISBN: 0-8247-6004-2

PRINTED IN THE UNITED STATES OF AMERICA

PREFACE

This volume attempts to collect some of the more important and useful techniques used in cyclic nucleotide research (particularly cyclic 3',5'-adenosine monophosphate), and to present these techniques in so far as possible in such a form that each chapter may be used as a laboratory manual. The abbreviations "cyclic AMP" and "cyclic GMP" will be used throughout this volume to refer to the 3',5'-cyclic monophosphates of adenosine and guanosine, respectively.

In any volume such as this, there will be some degree of crossover between chapters; such overlap may prove useful in this case. For instance, any of the assays for cyclic AMP may be used for measuring the rate of the adenylate cyclase reaction, or protein kinases specifically activated by either cyclic AMP or cyclic GMP have been useful in the assay of these nucleotides. Although two chapters describe preparation of isolated lipocytes, both are included because the preparation techniques differ according to the use for which the cells are intended.

The volume is divided into three basic parts; Assay of Cyclic Nucleotides, Enzymology of Cyclic Nucleotides, and The Use of Intact Cell Systems. The first part includes descriptions of tissue preparation and five different assays for cyclic AMP (and cyclic GMP). The second part includes preparation of liver plasma membranes and lipocyte "ghosts" and assay of adenylate cyclase in these preparations, and the preparation and assay of guanylate cyclase, cyclic nucleotide phosphodiesterase and cyclic AMP- and cyclic GMP-dependent protein kinases. The third part describes two whole cell systems, isolated lipocytes and isolated adrenal cells, and also includes the techniques of prelabeling pools of ATP with radioactive precursors in a variety of tissues.

Princeton, New Jersey
July, 1972

Mark Chasin

CONTRIBUTORS TO THIS VOLUME

GARY BROOKER, Department of Pharmacology, University of Virginia, School of Medicine, Charlottesville, Virginia 22901

MARK CHASIN, Department of Biochemical Pharmacology, Squibb Institute for Medical Research, New Brunswick, New Jersey 08903

JOHN DALY, National Institute of Arthritis and Metabolic Diseases, National Institute of Health, Bethesda, Maryland 20014

CHARLES A. FREE, Department of Biochemical Pharmacology, Squibb Institute for Medical Research, New Brunswick, New Jersey 08903

DON N. HARRIS, Department of Biochemical Pharmacology, Squibb Institute for Medical Research, New Brunswick, New Jersey 08903

ROGER A. JOHNSON, Department of Physiology, Vanderbilt University, Nashville, Tennessee 37232

JYH - FA KUO, Department of Pharmacology, Yale University School of Medicine, New Haven, Connecticut 06510

SHARON J. NORTHUP, Space Sciences Research Center and Department of Biochemistry, University of Missouri, Columbia, Missouri 65201

MARTIN RODBELL, Section on Membrane Regualtions, National Institute of Arthritis and Metabolic Diseases, Bethesda, Maryland 20014

IRA WEINRYB, Department of Biochemical Pharmacology, Squibb Institute for Medical Research, New Brunswick, New Jersey 08903

ARNOLD A. WHITE, Space Sciences Research Center and Department of Biochemistry, University of Missouri, Columbia, Missouri 65201

TERRY V. ZENSER, Space Sciences Research Center and Department of Biochemistry, University of Missouri, Columbia, Missouri 65201

CONTENTS

Chapter 3. HIGH-PRESSURE LIQUID CHROMATOGRAPHY FOR THE ANALYSIS OF CYCLIC NUCLEOTIDES

Gary Brooker

PART 2. ENZYMOLOGY OF CYCLIC NUCLEOTIDES

Chapter 4. METHODS FOR THE ISOLATION OF RAT LIVER PLASMA MEMBRANES AND FAT CELL "GHOSTS"; AN ASSAY METHOD FOR ADENYLATE CYCLASE

Martin Rodbell

Methods in

CYCLIC NUCLEOTIDE RESEARCH

Part 1

ASSAY OF CYCLIC NUCLEOTIDES

Chapter 1

ASSAY OF CYCLIC AMP

Roger A. Johnson

Department of Physiology
Vanderbilt University
Nashville, Tennessee 37232

From its outset, the study of the biological role of cyclic AMP has been greatly hampered by inadequacies in methodology. The most significant problems have been the very low concentration of this nucleotide in tissue and the rapidity with which its concentration within cells can change, whether being elevated or decreased. For these reasons the fixation and preparation of samples for assay are as critical as the analytical procedure for its determination. In both areas there are significant pitfalls. In this chapter, we shall describe some of the procedures that we have found to be particularly useful for the preparation and analysis of cyclic AMP samples.

I. SAMPLE PREPARATION

Several of the procedures used in the preparation of samples may depend very much on the method of analysis to be used (see Chapter 2, Section I.G). However, in general, sample treatment involves the following steps: fixation, extraction, fractionation and analysis. These steps will be described in this section. Some variations of these steps peculiar to a given assay will be discussed in greater detail in the section on the particular assay. A general scheme for sample preparation is shown in Fig. 1.

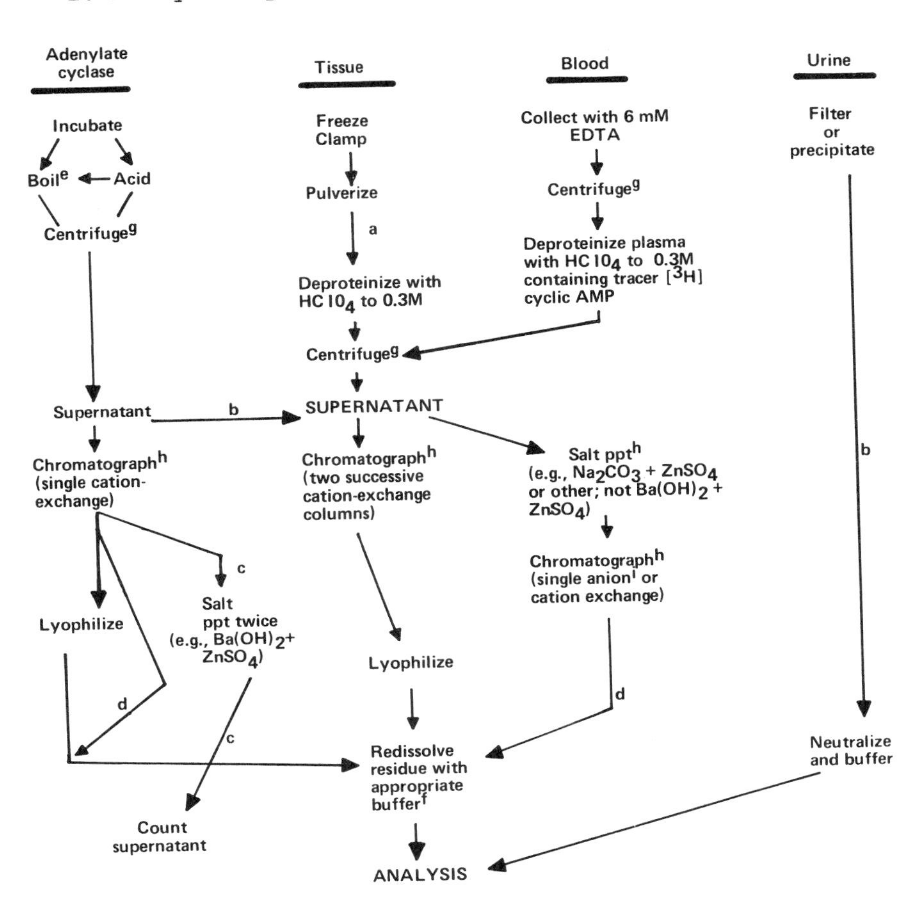

Fig. 1. Scheme for the preparation of samples for cyclic AMP analysis.
[a]tracer [^{3}H]-cyclic AMP added. [b]for luciferase-luminescence assay for cyclic AMP. [c]this sequence can only be used in conjunction with the use of labeled substrate (ATP). [d]lyophilization of eluates can sometimes be avoided if sample cyclic AMP content is high or if

the elution medium is adjusted to elute cyclic AMP in a smaller fraction (e. g., elute with water or buffer). [e]not useful at neutral pH if 5-10 mM Mn^{2+} is present in the incubation. [f]for the luminescence assay, 50 mM glycyl-glycine, pH 7.5; for the protein binding assay, 50 mM Na-acetate-acetic acid, pH 4.0; for the phosphorylase activation assay, 10 mM tris-HCl, pH 7.4. [g]centrifugal force need only be sufficient to sediment denatured protein, etc. [h]see text. [i]if an anion-exchange column is used, 1 mM EDTA should be included in the assay for either the luciferase or binding assay.

A. Fixation and Extraction

Inasmuch as cyclic AMP in tissue has a very high turnover rate, the method of sample fixation is especially critical in tissue studies. Perhaps the most satisfactory procedure for tissue fixation is very rapid freezing (for other methods, see Chapter 2, Section I.G.1). At the appropriate time, the tissue specimen is clamped between relatively massive metal blocks previously cooled in liquid nitrogen. The frozen tissue is then pulverized in a stainless steel impaction mortar, also at liquid nitrogen temperature. Various procedures have been successfully used for the further fixation and extraction of cyclic AMP from tissue. We have found the following method quite satisfactory. An aliquot of the tissue powder is weighed and then homogenized as rapidly as possible in a motor-driven glass-Teflon homogenizer with about 10 volumes of ice-cold 0.3 M perchloric acid containing about 5000-10000 cpm of tracer [^{3}H]-cyclic AMP per sample for the estimation of cyclic AMP recovery. Trichloroacetic acid and HCl have also been used satisfactorily. The specific activity of the [^{3}H]-cyclic AMP should be as high as possible to preclude assay complications. [^{3}H]-cyclic AMP of about 24 Ci/mmole is comercially available and is adequate for most purposes.

Cyclic AMP in plasma can readily be determined if blood is handled in the following way. Human blood is capable of metabolizing cyclic AMP with an apparent half-life ranging from 30 to 150 min at 37°C [1]. [^{3}H]-cyclic AMP is degraded much more rapidly in rat blood [2] and additional care is necessary. This degradation can be minimized by the reduction of temperature and the addition of methylxanthines or EDTA, and is presumably owing to phosphodiesterase activity. Samples of whole blood are drawn into EDTA (final concentration 6 mM) [3], mixed and centrifuged at 6000 x g for 5 min at 4°C. Tracer [^{3}H]-cyclic AMP (5000-10,000 cpm per sample) is added to an aliquot of plasma, to which perchloric acid is then added, to a final concentration of 0.3 M. Denatured protein, etc., is removed by centrifugation. Deproteinized plasma can be stored as such at 4°C for up to three months without measurable loss of the cyclic nucleotide. Prior to assay, the plasma is chromatographed as described in subsection B.

Assays of cyclic AMP are required also to determine adenylate cyclase activity. In this case, the enzyme reaction is generally terminated by the addition of acid (e.g., HCl to a concentration of 0.1 N), or the sample is placed in boiling water for 2 to 5 min, or both. If adenylate cyclase is being determined by measurement of cyclic AMP formed from unlabeled substrate, it will also be necessary to add tracer [^{3}H]-cyclic AMP to the sample at the time the reaction is terminated so that the cyclic AMP recovery can be estimated (see preceding paragraph). Some further caution should be exercised at this stage. If a solution containing ATP is heated at a neutral or high pH in the presence of certain metals, for example Ba^{2+} [4] or Mn^{2+} [5], cyclic AMP can be formed nonenzymatically. This problem has not been observed with Mg^{2+}, or at acid pH.

B. Fractionation

Depending on which assay for cyclic AMP is to be used, the extracted, deproteinized sample containing cyclic AMP will have to be further fractionated to a greater or lesser degree. A relatively high degree of sample purity is required for the luciferase-luminescence assay (see next paragraph) while comparatively crude samples can be assayed by the cyclic AMP-protein binding assay ([6]; see also Chapter 2) or by radioimmunoassay (see Chapter 2).

Samples may be further purified by several chromatographic procedures [7,8]. For the luciferase-luminescence assay, we have found it convenient to use cation-exchange chromatography. The sample is applied to a column (Bio-Rad AG50-X8, 100-200 mesh, H^+ form) previously equilibrated with 0.1 N HCl; the sample is also eluted with 0.1 N HCl. A sample of 5 ml may be applied to a column 30 x 0.6 cm; smaller samples would require shorter columns (e.g., 2.5 ml, 20 x 0.6 cm; 1-1.5 ml, 10 x 6 cm). The fraction containing cyclic AMP is lyophilized, the residue is redissolved in 0.1 N HCl and is applied to a second column, usually 20 x 0.6 cm, and it is chromatographed as before. The eluate from the second column is then lyophilized, redissolved in 50 mM glycyl-glycine buffer, pH 7.5, and assayed as described in Section II. The degree of purity required and, therefore, the size of the columns used, will depend greatly on the amount of cyclic AMP in the sample. Samples containing high concentrations of cyclic AMP will generally require less purification than samples of low cyclic AMP content

An alternative procedure to the preceding is to first precipitate the noncyclic adenine nucleotides by salt coprecipitation [9,10]. For example, to a 1.2-ml sample from an adenylate cyclase incubation are added 0.10 ml each of 1 M Na_2CO_3 and $ZnSO_4$ [11]. The salt combination of $Ba(OH)_2$ and $ZnSO_4$ cannot be used satisfactorily for this step without prior chromatography of the sample, because of the problem previously described regarding the nonenzymatic formation of cyclic AMP. The supernatant fraction resulting from this precipitation step can then be further purified by ion-exchange chromatography. This additional chromatography step is necessary to remove the Zn^{2+} as well as to further separate unprecipitated noncyclic adenine nucleotides from cyclic AMP. Each of these substances could interfere with the assay of cyclic AMP.

In our use of the protein-binding assay for cyclic AMP [6] we have found it necessary to chromatograph samples on at least one cation-exchange column prior to assay. A 10 x 0.7-cm column (resin, as described in Section I.B) has been used satisfactorily for samples up to 5 ml. Further discussion may be found in Chapter 2. Without such chromatography, incorrect cyclic AMP values were obtained, presumably owing to interfering agents that also competed for [^{3}H]-cyclic AMP binding to the protein. These discrepancies varied from tissue to tissue. For these and other obvious reasons, it is strongly suggested that, regardless of which assay for cyclic AMP is used, it be rigorously established that one is indeed measuring only cyclic AMP. This should be done by several procedures. First, samples should be treated with excess 3', 5'-nucleotide phosphodiesterase, to demonstrate the loss of assayable cyclic AMP and to provide an estimation of the degree of contamination or interference. Second, samples can be assayed in the presence and absence of known standards added to the samples, to establish whether interfering substances may be present. Third, some samples should be rigorously purified, to establish if a shorter, less exacting purification technique is adequate for the intended purposes. The use of the first two of these measures is described for the luciferase-luminescence assay for cyclic AMP.

The cation-exchange resin used in the preceding chromatography steps should be thoroughly washed prior to use. If anion-exchange resin (Bio-Rad AG2-X8, 100-200 mesh, Cl^- form) is used as an alternative purification step (cf. Figure 1), it can be washed in a similar manner. The resin is placed in a large column about 5-10 cm in diam). The following amounts of reagents are successively applied per pound of resin: (1) 3 liters of 0.5 N NaOH; (2) 3 liters of distilled water; (3) 4 liters of 2 N HCl; (4) 3 liters of distilled water; and (5) 1.5 liters of 0.1 N HCl. The resin is then stored in a refrigerator. This column procedure is preferable to batch washing in that it is faster, more efficient, and more effective.

The preparation of urine samples for assay by luciferase-luminescence is straightforward. They need only to be filtered or otherwise freed of sediment, and neutralized prior to assay.

II. LUCIFERASE-LUMINESCENCE ASSAY OF CYCLIC AMP [12, 13]

A. Reaction Sequence

The luminescence assay of cyclic AMP is based on

its conversion to ATP and the subsequent determination of ATP by its luminescent reaction with firefly luciferin and luciferase. Cyclic AMP is converted to ATP and then determined according to the following reaction sequence:

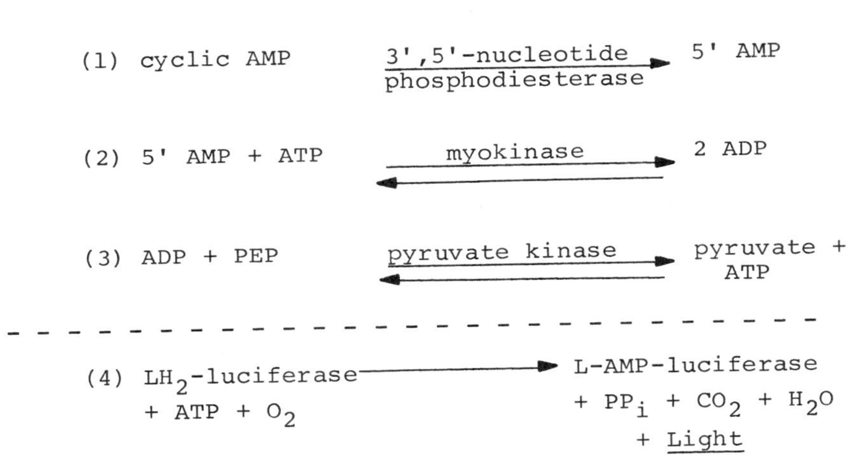

In reaction (4), LH_2 refers to the reduced form of luciferin, a substrate for luciferase; the CO_2 and H_2O are derived from LH_2 upon oxidation to L-AMP-luciferase. It is apparent from reactions (1)-(3) that AMP or ADP present in the sample or incubation mixture will also be converted to ATP and react with luciferin and luciferase. Thus it is necessary for these interfering agents to be removed from the sample prior to assay, as described previously.

Further, to correct for small amounts of contamination remaining after purification, each sample is assayed in the presence and absence of excess phosphodiesterase

B. Assay Procedure

The assay is carried out in two stages, in a single disposable culture tube (6 x 50 mm); the first stage consists of reactions (1)-(3), then a second stage is reaction (4). The conversion of cyclic AMP to ATP involves incubation of the purified sample with the enzymes 3',5'-nucleotide phosphodiesterase [14], myokinase, and pyruvate kinase, all of which are available from commercial sources. (Some preparations of commercial phosphodiesterase may be contaminated with excessive nucleotidase and warrant further purification prior to use [14]). Each sample is assayed in duplicate, with and without phosphodiesterase, and with phosphodiesterase plus an added standard. This method corrects, respectively, for the presence of contaminating noncyclic adenine nucleotides and for potential assay inhibition by unknown substances. To 25-μl sample aliquot, with a concentration of cycli AMP between 10^{-8}M and 10^{-5}M, is added 50 μl of reagent containing: 10 mM $MgSO_4$; 100 mM glycyl-glycine buffer, pH 7.5; 0.1 mM phosphoenolpyruvate (PEP); 10^{-10} M ATP:

5mM dithiothreitol [15]; 2 μg myokinase; 10 μg pyruvate kinase; and, when used, 10 mU phosphodiesterase. The volume is made up to 100 μl with either 25 μl water or with a solution of an external standard. The resulting solution is then capped and incubated overnight at 30°C or, alternatively, for 2 hr at 37°C, whichever is more convenient. The incubated samples can be kept on ice for immediate determination of ATP or frozen for later analysis.

The determination of the ATP by luminescence is based on reaction (4), which is essentially complete in 1 sec. The peak intensity of light produced is directly proportional to the initial ATP concentration (see Fig. 2). After the initial peak, the light emission decays by a process dependent on numerous factors. Because of the nature of the reaction, it is desirable, but not necessary, to measure the initial peak light intensity. Any method would be satisfactory in which rapid mixing of the ATP and luciferase is achieved, and in which the reaction is monitored by a sensitive photomultiplier. Any of several instruments may be used for this purpose. A fluorometer, with light source and filters removed, may be used in conjunction with a short rise-time recorder, and the peak light intensity may be read

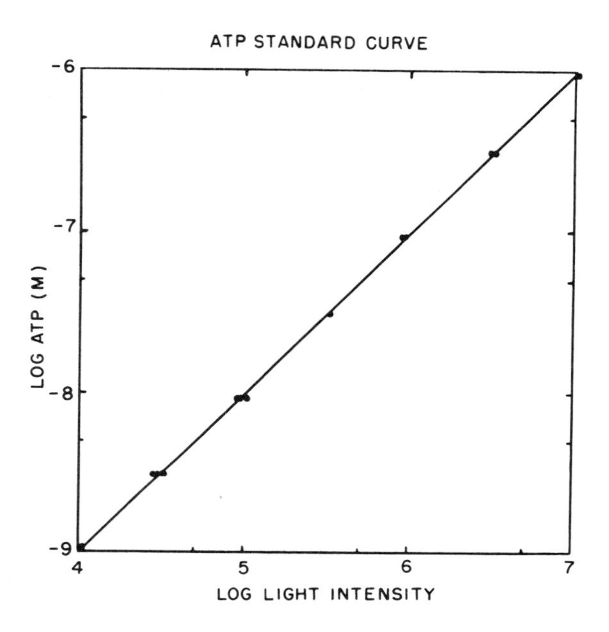

Fig. 2. The luciferase-luminescence standard curve for ATP. Values for the logarithm of light intensity are based on an arbitrary scale. Taken from R. A. Johnson [13].

from the recorder. A liquid scintillation counter may also be used [16]. However, in this case it is technically difficult to record the peak light intensity, and a small area under the curve is usually counted past the peak. In the absence of interference by substances that change the shape of the light decay curve, the count will be proportional to the initial ATP concentration. Because of the recording problem, it is important that the "counting period" for each sample begin at precisely the same time relative to the mixing of luciferase with ATP. In our hands, the most convenient of the alternative monitoring devices has been the Luminescence Biometer (from E. I. DuPont de Nemours and Co., Inc., Instrument Products Division), designed to electronically store the peak light intensity, which can be released later for digital readout. The following description of the procedure for measurement of ATP is, therefore, based on the use of such an instrument.

The incubation tubes in which ATP is generated also serve as cuvettes for the Luminescence Biometer. The cuvette is inserted into the instrument directly in front of the photomultiplier. A 20-µl aliquot of luciferin-luciferase reagent is then injected into the tube by means of a 50-µl syringe (Hamilton), to which has been attached a spring-loaded device (Repro-Jector, from Shandon Scientific Co., Sewickley, Pa.) that expels the syringe contents at a reproducible velocity. The luciferin-luciferase reagent contains: 100 mM glycyl-glycine buffer, pH 7.5; 10mM $MgSO_4$; 3 mM dithiothreitol; 1 mg/ml bovine serum albumin;

and 20 mg/ml luciferin-luciferase complex. Crystalline luciferase and luciferin have been prepared as a stable enzyme-substrate complex, which is available from DuPont. Unsatisfactory results will be obtained if cruder preparations of luciferase are used. Inclusion of bovine serum albumin and dithiothreitol is necessary to stabilize the luciferase. The injection velocity is such that adequate mixing results. The peak light intensity can be read from the instrument.

The luciferase-luminescence reaction for ATP is an extremely sensitive and rapid procedure. It is optimally sensitive to about 2 x 10^{-16} mole of ATP, and is linear over five orders of magnitude of concentrations of the nucleotide. The range of the assay can be readily adjusted by changing either the luciferase-luciferin concentration or the photomultiplier sensitivity. A typical ATP standard curve under conditions commonly used in the assay of cyclic AMP is shown in Figure 2.

The standard curve for the determination of cyclic AMP by the luminescence assay is shown in Figure 3. The cyclic AMP standards were incubated in the absence and presence of phosphodiesterase, as represented by the two solid lines in the figure. The dashed line

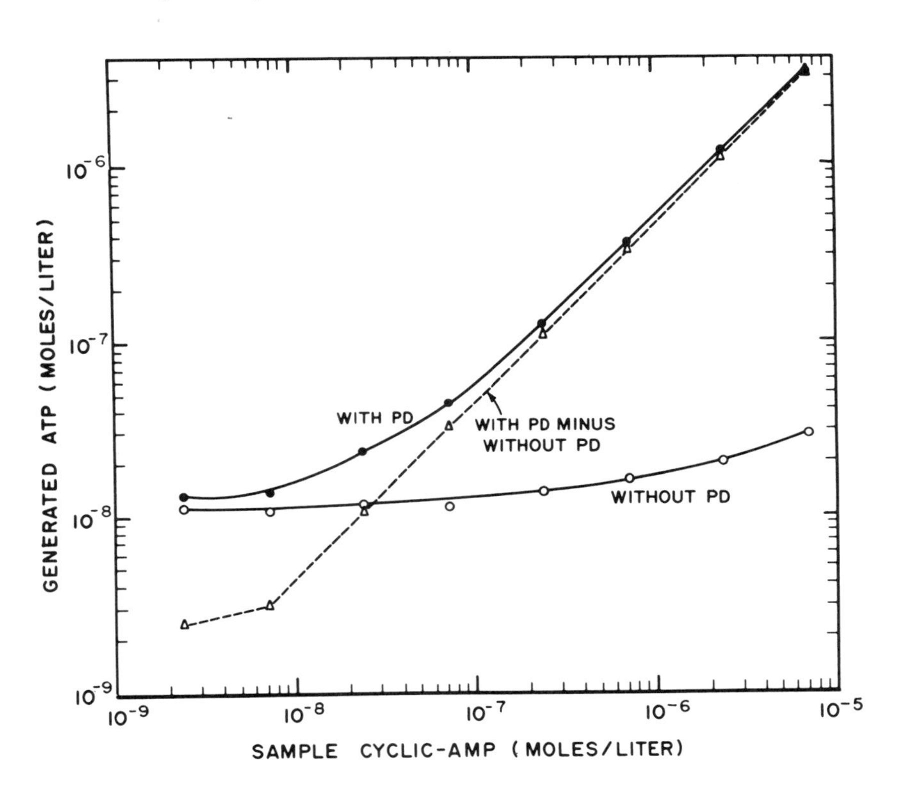

Fig. 3. Standard curve for the assay of cyclic AMP by the luciferase-luminescence assay. Cyclic AMP concentrations are those in the standard solution, from which 100 μl aliquots were taken. ATP concentrations are those in a final 200 μl incubation mixture. PD refers to 3',5'-cyclic nucleotide phosphodiesterase. Myokinase and pyruvate kinase were treated with cocoanut charcoal before use in the incubation mixture. From Johnson et al. [12].

represents the difference between these two curves and, therefore, the ATP generated from cyclic AMP. The standard curve was linear from 7.2×10^{-9} to 7.2×10^{-6} M cyclic AMP, equivalent to 7.2×10^{-13} to 7.2×10^{-10} mole in the 100-μl sample used in this experiment. The conversion of cyclic AMP to ATP was about 90% complete. In this experiment the assay blank was equivalent to about 2.7 pmole of cyclic AMP, owing principally to contamination by nucleotides bound to the enzymes used to convert cyclic AMP to ATP. The most contaminated of these is myokinase. The assay blank will vary depending on the extent of contamination in each particular batch of enzymes. Removal of these nucleotide contaminants markedly enhances the sensitivity of the assay. The contaminating nucleotides can be partially removed from the enzymes by chromatography on short charcoal and anion-exchange columns (0.48 x 5 or 10 cm; activated cocoanut charcoal, previously exhaustively washed, 50-200 mesh; Bio-Rad AG2-X8, 100-200 mesh, Cl^- form), equilibrated and eluted with 5 mM tris-HCl buffer, pH 7.3. The most effective procedure used both methods successively. The assay blank was sufficiently reduced to permit the detection of 0.05 pmole of cyclic AMP.

1. Assay of Cyclic AMP

Dr. Charles A. Sutherland, working in this laboratory, has been using Sephadex QAE (A-25 from Pharmacia) to chromatograph and decontaminate myokinase and pyruvate kinase. The enzymes are incubated in the presence of phosphoenolpyruvate (PEP) under the conditions described for reactions (2) and (3) before being applied to the column, previously equilibrated with PEP. Contaminants such as AMP and ADP are converted to ATP, thereby facilitating the adsorption of nucleotide to the anion-exchange matrix. By this procedure, the assay blank can be significantly reduced; this procedure has made the assay of samples containing 0.1 pmole of cyclic AMP, or less, more practical.

C. Assay Advantages and Disadvantages

The luminescence assay for cyclic AMP exhibits linearity over three orders of magnitude, with a sensitivity to about 1 pmole of cyclic AMP. Assay sensitivity can be increased tenfold by removal of nucleotide contaminants from myokinase and pyruvate kinase as described in subsection B, thereby increasing the linear range to four orders of magnitude and the sensitivity to 0.1 pmole of cyclic AMP. The assay specificity is assured by a three-stage discrimination. The first stage of discrimination is the nucleotide precipitation and cation-exchange chromatography.

Almost all contaminants are eliminated by these steps. The second stage is the incubation with phosphodiesterase which is specific for the 3',5'-cyclic phosphate moiety. The third stage is the incubation with luciferase, which is specific for the base moiety of ATP. The procedure is reproducible, generally ±5%, and relatively rapid, permitting the assay of approximately 150-200 samples per man-week.

Inasmuch as the assay is based on the determination of ATP, a further advantage of the assay is the relative ease with which it could be extended to the measurement of other substances, e.g., nucleotides convertible to ATP. Analogously, the activity of enzymes either that utilize ATP or whose products are convertible to ATP could readily be determined. Such assays for other substances also could be extremely sensitive and rapid, and would share many of the advantages of the assay of cyclic AMP described herein.

There are certain disadvantages to the luminescence assay of cyclic AMP. The most important of these is the necessity to remove essentially all noncyclic adenine nucleotides from samples and, for maximun assay sensitivity, to remove them from the enzymes used in the assay as well. The second disadvantage is the cost of instrumentation. Although the assay may be performed with a fluorometer or scintillation counter, the most convenient instrument to use has been the Luminescence Biometer.

III. PHOSPHORYLASE ACTIVATION ASSAY FOR CYCLIC AMP [17]

The first assay for cyclic AMP naturally evolved from the experiments that led to its discovery, and was based on the ability of cyclic AMP to enhance the rate of activation of inactive liver phosphorylase (ILP). A description of this procedure is included here to permit the reader to view the newer assays in a historical context, with some degree of comparative insight. Since its initial description the assay has undergone a number of minor modification, taking place over a period of years and reported in several publications. The following description of the assay will take most of these modifications into account, but no attempt will be made to describe the preparation of certain reagents required for the assay of cyclic AMP by this method; these have recently been compiled and published elsewhere [18].

A. Reaction Sequence

With our current understanding of the role of cyclic AMP to modulate glycogen metabolism, the assay procedure as originally described might be translated into the following schematic reaction sequence:

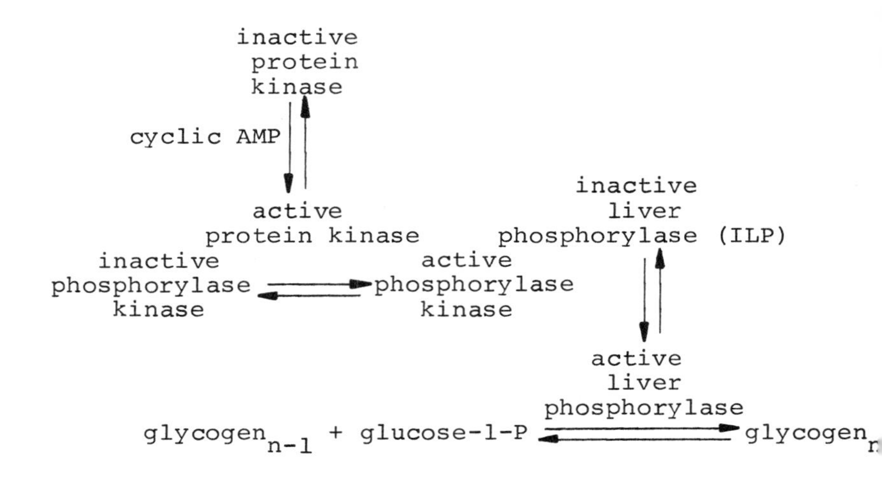

The assay is carried out in three stages: (1) ILP is converted to the active enzyme at a rate dependent on the initial cyclic AMP concentration; (2) the amount of active liver phosphorylase is determined by its ability to catalyze the conversion of glycogen$_{n-1}$ and glucose-1-phosphate to glycogen$_n$; and (3) glycogen is determined by its colorimetric reaction with iodine.

B. Assay Procedure

Each standard or blank is assayed in duplicate, and samples, prepared as described previously are usually assayed at two dilutions, each in duplicate. A 20-μl aliquot of sample, standard, or blank is pipetted into a 10 x 75-mm disposable culture tube, to which is then added 20 μl of reagent containing

200 mM tris-HCl buffer, pH 7.4; 15 mM $MgSO_4$; 12 mM ATP; 5.0 mM caffeine; and 0.6 mM EDTA. Standards are made up in 10 mM tris-HCl buffer, pH 7.4, which also constitutes the assay blanks and is used to dissolve the residue from lyophilized samples. The standards routinely used range between 0.1 and 6.0 pmole of cyclic AMP. To each of the preceding tubes is added 20 µl of enzyme reagent, at timed intervals, containing from 66 to 266 µg of purified glycogen, 0.02 to 0.04 unit of ILP, 1 to 5 µl of an 11,000 x g supernatant fraction of a 33% homogenate of dog liver, and 24 µM l-epinephrine in 10 mM tris-HCl buffer, pH 7.4. The supernatant fraction from dog liver contains both the protein kinase and the phosphorylase kinase previously described. The amounts of reagents in this second addition tend to vary, since each supernatant fraction must be empirically balanced with optimal amounts of ILP and glycogen. These tubes are then kept in an ice-water bath for exactly 30 min before being transferred to a 25°C water bath for 20 min, also at timed intervals. During this incubation the ILP is converted to the active enzyme at a rate dependent on the initial cyclic AMP concentration.

The second stage of the assay is initiated by the addition to each tube of 200 µl of a reaction mixture, pH 6.1, containing 2.2 mM 5'-AMP, 45 mM glucose-1-phosphate, 120 mM NaF, and 1.0 mg of glycogen, and the transfer of the tubes directly to a 37°C water bath. This reaction converts $glycogen_{n-1}$ to $glycogen_n$. After a 30-min incubation, 200 μl aliquots are removed from each sample and placed into tubes containing 1.0-1.5 mg of I_2, 2.0-3.0 mg of KI, and 0.04 meq of HCl in 1.0 ml to terminate the second-stage reaction and permit determination of the glycogen formed. This mixture is diluted with 8.0 ml of distilled water, and the amount of glycogen-iodine complex is determined by measurement of its absorbance at 540 nm, with distilled water as reference. Unknown cyclic AMP values are obtained from the standard curve, which is a plot of absorbance (less the tris-HCl blanks) against the concentration of known cyclic AMP standards.

C. Assay Characteristics and Limitations

The assay is sensitive to about 0.1 pmole of cyclic AMP, although it is useful only over a relatively narrow concentration range. The volume of the assay can be easily reduced to conserve certain of the reagents, which are somewhat time consuming to prepare, although in our hands it has been convenient to perform this assay as described. The sensitivity of the assay can be further increased by the use of labeled glucose-1-phosphate (about 0.05 μCi of [U-^{14}C]-glucose-1-P per sample), and the determination of labeled glycogen [19]. Under appropriate conditions, 0.03 pmole of cyclic AMP could be detected.

The assay has several disadvantages. Perhaps the most obvious drawback is that the enzymes required for the assay are not commercially available and must be prepared, and that several of the reagents require special attention [18]. Although the assay displays certain quantitative variability, i.e., the standard curve may vary from one assay to the next, the values obtained with it are in good agreement with values obtained using the more recently developed analytical methods. A further disadvantage is that a large number of agents can interfere with the assay. Although most of these agents are removed by the chromatographic procedures, apparently certain elusive unknown factor(s) may still be present [20,21]. Even though the phosphorylase activation assay may be difficult to set up and sometimes cumbersome to use and maintain, and perhaps requires some artisan as well as scientific ability, it is and has been used successfully by a large number of investigators to provide the foundation of our current understanding of the biological role of cyclic AMP.

REFERENCES

[1] A. E. Broadus, N. I. Kaminsky, J. G. Hardman, E. W. Sutherland, and G. W. Liddle, J. Clin. Invest., 49, 2222 (1970).

[2] W. D. Patterson, J. G. Hardman, and E. W. Sutherland, Fed. Proc., 30, 220 Abs. (1971).

[3] A. E. Broadus, J. G. Hardman, N. I. Kaminsky, J. H. Ball, E. W. Sutherland, and G. W. Liddle, Ann. N.Y. Acad. Sci., 185, 50 (1971).

[4] W. H. Cook, D. Lipkin, and R. Markham, J. Amer. Chem. Soc., 79, 3607 (1957).

[5] J. P. Gray, Ph.D. Dissertation, Vanderbilt University (1971).

[6] A. G. Gilman, Proc. Natl. Acad. Sci., U. S., 67, 305 (1970).

[7] E. Ishikawa, S. Ishikawa, J. W. Davis, and E. W. Sutherland, J. Biol. Chem., 244, 6371 (1969).

[8] N. D. Goldberg, J. Larner, H. Sasko, and A. G. O'Toole, Anal. Biochem., 28, 523 (1969).

[9] G. Krishna, B. Weiss, and B. B. Brodie, J. Pharmacol. Exp. Ther., 163, 379 (1968).

[10] P. S. Chan, C. T. Black, and B. J. Williams, Fed. Proc., 29, 616 Abs. (1970).

[11] M. C. Lin, Personal communication (1970).

[12] R. A. Johnson, J. G. Hardman, A. E. Broadus, and E. W. Sutherland, Anal. Biochem., 35, 91 (1970).

[13] R. A. Johnson, in Advances in Cyclic Nucleotide Research, Vol. 1, p. , (P. Greengard, R. Paoletti, and G. A. Robison, eds.), Raven Press, New York, 1972.

[14] R. W. Butcher and E. W. Sutherland, J. Biol. Chem., 237, 1244 (1962).

[15] C. A. Sutherland, Unpublished observations.

[16] M. S. Ebadi, B. Weiss, and E. Costa, Science, 170, 188 (1970).

[17] T. W. Rall and E. W. Sutherland, J. Biol. Chem., 232, 1065 (1958).

[18] G. A. Robison, R. W. Butcher, and E. W. Sutherland, Cyclic AMP, Academic Press, New York, 1971.

[19] R. J. Ho, Personal communication (1970).

[20] R. W. Butcher, E. W. Sutherland, and T. W. Rall, Pharmacologist, 2, 66 (1960).

[21] F. Murad, T. W. Rall, and M. Vaughan, Biochim. Biophys. Acta, 192, 430 (1969).

Chapter 2

PROTEIN BINDING ASSAYS FOR CYCLIC AMP: RADIOIMMUNOASSAY AND CYCLIC AMP-DEPENDENT PROTEIN KINASE BINDING ASSAY

Ira Weinryb

Department of Biochemical Pharmacology
Squibb Institute for Medical Research
New Brunswick, New Jersey 08903

I. INTRODUCTION

Several approaches to the measurement of adenosine 3',5' cyclic monophosphate and other cyclic nucleotides in biological materials are found in the literature, and brief discussions of the nature and relative merits of many of the methods are available [1,2]. This chapter covers two of the newer assay techniques, the radioimmunoassay devised by Steiner et al. [3] and the cyclic AMP-dependent protein kinase binding assay developed by Gilman [4]. These methods are of particular interest because of their high sensitivity and specificity. They are both examples of a "competitive protein-binding saturation assay," in which a binding protein (e.g., antibody or kinase) is saturated with the labeled compound of interest (e.g., [^{3}H]-cyclic AMP), and this pool of marker is then diluted with an aliquot of unlabeled compound, of either known (standard) or unknown (e.g., urine sample or tissue extract) content. Competition between labeled and unlabeled compound results in less radioactivity being bound to the protein;

the decrease can be translated into the amount of unlabeled compound added in the aliquot with the aid of a standard curve. Other protein-binding assays for cyclic AMP [5,6] and guanosine 3'5' cyclic monophosphate [7] have been reported, but will not be considered here in detail.

II. RADIOIMMUNOASSSAY

The radioimmunoassay procedures carried out in our laboratory are modeled after those published by Steiner and colleagues [3]. Alternatives to, or variations of, the standard methodology will be included in the following description of the assay. The basic steps in the development of this assay are as follows: (1) cyclic AMP (or cyclic GMP) is rendered immunogenic by conjugation to a (large molecular weight) carrier protein; (2) animals are immunized with the conjugate, and develop antibodies to the nucleotide moiety; (3) bleedings are collected and antisera are isolated; and (4) aliquots of suitably diluted antisera or globulin fractions derived therefrom are used in a saturation binding assay of the type described in Section I.

A. Preparation of Cyclic Nucleotide-Protein Conjugate

Preparation of the conjugate involves two distinct steps: (1) alteration of the nucleotide to provide a locus for conjugation, and (2) coupling of the nucleotide derivative to the carrier macromolecule. In this assay, cyclic AMP (or cyclic GMP) is succinylated at the 2'O position of the ribose ring, and the free carboxyl group remaining is coupled to a protein carrier, presumably at loci such as ε-amino functions of lysine residues. The following procedures deal explicitly with the assay for cyclic AMP.

1. Synthesis of 2'O-Succinyl Cyclic AMP

Cyclic AMP, 230 mg (0.7 mmole), and 4'-morpholine N,N'-dicyclohexylcarboxyamidine (MDCC)[1], 222 mg (0.76 mmole), are dissolved in 7.5 ml of hot anhydrous pyridine (Redistilled or other high quality (chromatographic or spectroscopic grade) pyridine dried over solid KOH is recommended.) in a reaction vessel fitted with a reflux condenser and a drying tube. The reaction mixture is allowed to cool, and 3.75 g or less (1.0 g is usable [13]) of succinic anhydride (37 mmoles) is added. The suspension is stirred at room temperature for 18 hr (overnight) in the absence of moisture. The reaction mixture turns from pale yellow in appearance to dark and opaque by the end of the 18-hr period. Then 3.75 ml water is added[2] and the mixture is left at 4°C for 5 hr. Inspection reveals a whitish precipitate. Rotary evaporation at 40°C removes the pyridine, leaving a light-colored residue, which is redissolved with stirring in 10 ml of water. Adjustment of the pH to 2.0 with 5 N HCl causes the solution to turn turbid and viscous. This slurry is layered atop a 2.5 x 28 cm column of Dowex AG 50W-X8, 100-200 mesh, H^+ form, and the column is eluted with distilled water. Fractions are collected (5 ml) and monitored by the absorbance at 260 nm and by thin-layer chromatography (TLC) on cellulose sheets (with fluorescent indicator) with a 95% ethanol-1.0 M ammonium acetate (3 mM in EDTA; pH 5.0), 7:3 (v/v), solvent. Cyclic AMP travels with an R_f value of 0.40-0.43, whereas a slower spot (presumably 2'<u>O</u>-succinyl cyclic AMP) travels with an R_f value of 0.31-0.33. [Ethanol-0.5 M ammonium acetate, 5:2 (v/v) also has been used successfully [3]. With this solvent, R_f values for cyclic AMP and 2'<u>O</u>-succinyl cyclic AMP were 0.27 and 0.13, respectively]. Two broad peaks containing UV-absorbing material are eluted from the column, comprising, approximately, tubes 10-40 and 60-120. The slower-moving spot on TLC resides mainly in the early portion of the second peak (tubes 60-120), whereas the spot traveling as cyclic AMP is concentrated in tubes 130-180. The group of tubes containing succinyl cyclic AMP is pooled and lyophilized, leaving a white to off-white powder.[3]

The product[4] may be analyzed by IR and NMR spectroscopy. Infrared analysis reveals the presence of an esterified carboxyl group by its absorption at 1725 cm^{-1}. Quantitative NMR measurements indicate that up to 90% of the total nucleotide recovered was the 2'O-succinyl derivative, although a variable amount of succinic acid may also be present [33]. Although this level of purity is apparently satisfactory, further purification has been reported via preparation of the barium salt [3,10].

2. Coupling of 2'O-Succinyl Cyclic AMP to Carrier

a. Human Serum Albumin as Carrier. The 2'O-succinyl cyclic AMP is coupled to human serum albumin (HSA) by the procedures of Steiner et al. [3]. To a 10-mg/ml solution of HSA (Nutritional Biochemical), crystallized four times in water (pH about 5.1) is added 5 mg/ml of the lyophilized succinyl cyclic AMP powder. The pH is adjusted with microliter amounts of 0.1 N NaOH to pH 5.5 ±0.2, and 5 mg/ml of 1-ethyl-3(3-dimethylaminopropyl) carbodiimide· HCL (EDC, Ott Chemical Co.) is added. The pH is again adjusted to

5.5, and the mixture is incubated overnight (16-20 hr) at room temperature in the dark. The product is dialyzed against phosphosaline buffer (0.15 M NaCl _plus_ 0.01 M NaH_2PO_4, pH 7.4) at 4°C for 24-48 hr with frequent changes of dialyzate. Spectral examination of the resulting conjugate reveals an absorption band at 260 nm, owing to nucleotide. The contribution of HSA to the spectrum of the conjugate is calculable from a spectrum of HSA alone at the appropriate concentration (the initial concentration, corrected for volume changes owing to pH adjustment and dialysis). The spectrum remaining is owing to bound nucleotide; its concentration can be estimated by assuming a molar extinction coefficient of 15,000. The ratio of bound nucleotide to HSA molar concentrations also represents the average number of cyclic AMP residues per HSA molecule (a molecular weight of 69,000 is assumed). We find an average of 4-5 residues of nucleotide per HSA molecule, slightly lower than the ratio of 5-6 reported elsewhere [3]. Substitution of cyclic AMP for the succinyl derivative in the coupling reaction results in binding of less than one nucleotide residue per HSA molecule, in agreement with the results of Steiner et al. [3]. The EDC does not affect the UV spectrum of HSA in the 250-280 nm region [3].

b. Poly-L-lysine as Carrier. Succinyl cyclic AMP may also be coupled to poly-L-lysine or Keyhole limpet hemocyanin [3]. We have used poly-L-lysine·HCl, with a molecular weight range of 75,000-200,000 (Schwartz/Mann), as the macromolecular species in the coupling reaction. Using conditions as described in Section II.A.2.a, about one residue of cyclic AMP is bound per 100 lysine residues. By doubling the nucleotide concentration, and increasing EDC to 7.5 mg/ml, the coupling ratio is raised to one residue of nucleotide per 60 lysine residues. These ratios are substantially higher than those obtained with HSA. Even higher nucleotide to lysine ratios might be attained if the coupling is done at higher pH values [a greater fraction of the ε-amino groups would be unprotonated (pK_a about 9.5)] if the random-coil-to-helix transition that occurs with the same pK_a value [12] does not render many of the residues inaccessible to the reacting nucleotide.

B. Labeled Cyclic AMP Marker

1. [^{3}H]-Cyclic AMP

[^{3}H]-cyclic AMP is a convenient marker for the radioimmunoassay. It has the following advantages: (1) commercial availability of a high purity supply, (2) relative radiochemical stability, (3) long half-life (12.3 yr), and (4) relatively low expense. In addition, [^{3}H]-cyclic AMP with a specific activity[5] as high as 24 Ci/mmole is available from New England Nuclear Co. Our experience has been confined to samples with specific activities of 14-17 Ci/mmole, purchased from Schwartz/Mann. The major disadvantage in the use of [^{3}H]-cyclic AMP as marker is that about 100 times as much of the antibody is necessary as if [^{125}I]-succinyl cyclic AMP, tyrosine methyl ester, of the same purity is used as marker to attain the same fraction of added counts bound to the antibody. A minor shortcoming is the somewhat decreased assay sensitivity encountered (1 pmole is the practical threshold). The use of [^{3}H]-cyclic AMP may well be considered, however, in situations where extensive or reliable supplies of antibody are at hand. The possibility exists that a more sensitive assay (that uses much less antibody) could be achieved by use of [^{3}H]-succinyl cyclic AMP, which shows about 200 times the affinity for the antibody that cyclic AMP does, as the marker [13].

2. [^{125}I]-Succinyl Cyclic AMP Tyrosine Methyl Ester

[^{125}I]-succinyl cyclic AMP, tyrosine methyl ester ([^{125}I]-SCAMP-TME), is the iodinated cyclic AMP derivative used in the development of the radioimmunoassay [3]. It can be synthesized with very high specific activities (>150 Ci/mmole), and permits immunoassay of cyclic AMP with a sensitivity well below 1 pmole, and with much more dilute antibody than is possible with [^{3}H]-cyclic AMP marker. The principal difficulties with this marker are its relative lability, its shorter half-life (60 days), and the relative difficulty of synthesis in high yield. The [^{125}I]-SCAMP-TME (as well as a complete radioimmunoassay kit) is available commercially from Collaborative Research (Waltham, Mass.) and from Schwartz/Mann (Orangeburg, N. Y.). Because of the instability of the marker (perhaps owing to its very high specific activity), it is prudent to determine the percentage of intact marker in any new batch by checking the percentage of radioactivity bound to increasingly concentrated solutions of antibody, until an apparent plateau has been reached.

The synthesis of this marker requires (1) coupling of tyrosine methyl ester to succinyl cyclic AMP, and (2) iodination of SCAMP-TME with ^{125}I.

a. Synthesis of SCAMP-TME. The original procedure [1,3] involved the use of N,N'-dicyclohexylcarbodiimide (DCC) as coupling agent. This procedure has been followed with the modification that triethylamine is not added if the free tyrosine methyl ester[6] is used [30]. Two equivalents of tyrosine methyl ester are added to one equivalent of succinyl cyclic AMP (free acid), which has been dissolved in a small volume of dimethylformamide (DMF) (0.1 ml/5-10 mg SCAMP). After the reaction mixture is cooled to -4°C in, for example, an ice-salt bath, one equivalent of DCC in cold DMF is added with stirring for 1 hr at -4°C and then for 48 hr at 4°C. The product may be purified by cellulose TLC [1,3] or paper chromatography [30]. The TLC solvent is ethanol-0.5 M ammonium acetate 5:2. The nitrosonaphthol-positive, ninhydrin-negative product (R_F = 0.8) may be eluted with 50% ethanol. Paper chromatographs are developed in butanol-acetic acid-H_2O 12:3:5. The product (R_F = 0.3) travels faster than unreacted SCAMP (R_F = 0.2) and a decomposition product, cyclic AMP (R_F = 0.1), and more slowly than the nitrosonaphthol-positive, ninhydrin-positive, unreacted tyrosine methyl ester (R_F = 0.5). The butanol-acetic acid-H_2O solvent is also suitable for analytical or preparative TLC [30]. Although yields of SCAMP-TME are relatively low (30-35%) with this method [30], such an objection is not overwhelming in view of the minute amounts used for iodination and in the assay.

Improved yields are reported, however, by the use of an alternative procedure that uses ethyl chloroformate in a mixed carboxylic-carbonic acid reaction [1,13]. This method has also been found suitable for the synthesis of the corresponding derivative of cyclic GMP. One equivalent of succinylated cyclic nucleotide and three equivalents of trioctylamine are dissolved in cold (0°C) DMF, and one equivalent of ethyl chloroformate in DMF is added. After 15 min at 0°C, two equivalents each of tyrosine methyl ester and trioctylamine in DMF are added, and the reaction is carried out an additional 18 hr at room temperature with continual stirring. The TLC of the reaction mixture with butanol-acetic acid-water 12:3:5 separates the desired product from unreacted SCAMP and tyrosine methyl ester. Yields of 40-55% are obtainable [13].

b. Iodination of SCAMP-TME. The SCAMP-TME may be iodinated [3] by the method of Hunter and Greenwood [27], with slight modifications [26]. Ten micrograms of SCAMP-TME in 50 μl of 0.5 M phosphate buffer, pH 7.5, are added to a 1-ml reaction vial, followed by 20 μl of carrier-free ^{125}I (4.8 mCi). The vial is stoppered and 50 μl of phosphate buffer containing 88 μg of chloramine-T is injected through the stopper.

The vial is shaken for 30 sec, and then 100 μl of sodium metabisulfate (240 μg) in phosphate buffer is added to terminate the reaction by reduction of residual iodine to iodide.

The [^{125}I]-SCAMP-TME may be purified by preparative paper chromatography [26]. The reaction mixture is spotted on ion-exchange resin-impregnated paper (WB2, from Reeve-Angel). Development is by descending chromatography, with a solvent of pyridine-ethanol-acetic acid-water 33:200:29:139. After 2 or 3 hr, three peaks of radioactivity are apparent. One, at the origin, is residual iodide. The second, at an R_F of 0.9, has not been identified but analogy to the results with other methods [3] indicates that it may be [^{125}I]-tyrosine methyl ester.

The iodinated cyclic nucleotide derivative may be purified also by chromatography on Sephadex G-10 [3], or by thin-layer chromatography on cellulose [13]. For TLC a solvent of butanol-glacial acetic acid-H_2O 12:3:5 is used [1]. This procedure is capable of separating [^{125}I]-SCAMP-TME (R_F = 0.6) from SCAMP-TME (R_F = 0.4), thereby increasing, if necessary the specific activity of the iodinated product obtained from preparative paper chromatography [23]. The reaction mixture may also be applied to a column, 0.9 cm x 16 cm, of Sephadex G-10 (a column 0.9 cm x 9 cm is actually sufficient [13]) previously washed with 1 ml of 3% HSA in phosphosaline buffer (Section II.A.2.a), pH 7.5. The column is eluted with phosphosaline buffer in 2-ml fractions; three radioactive peaks are again observed. The third peak (70-110 ml) is [^{125}I]-SCAMP-TME.

The iodinated marker has high specific activity (>150 Ci/mmole) and is stored in small aliquots at -20°C [3]. Specific activities >800 Ci/mmole have been recently obtained [13]. In such cases, less than 1 mg of cyclic nucleotide derivative was reacted with about 1 mCi of [^{125}I].

C. Immunization of Animals

Both rabbits (New Zealand white) and goats have been used as sources of antibody to cyclic AMP in our laboratories. Goats, with their considerably larger blood supply, represent more economical sources of antibody than rabbits, since both are immunized with similar amounts of the succinyl cyclic AMP-HSA conjugate.

Steiner et al. [1,3] have immunized randomly bred rabbits with 1 mg of conjugate suspended in Freund's complete adjuvant (0.25 mg/footpad). Animals were given booster injections of 0.2-0.6 mg of conjugate into two footpads at 6 week intervals, and were bled by cardiac puncture 10-14 days after booster injections. Adequate antibody titer has been reported after one booster injection.

We have immunized rabbits by various programs. Unless otherwise stated, 6-week intervals between injections have been used. In one scheme, 1 mg of conjugate in Freund's complete adjuvant (FCA) is injected into the footpads. A second injection of 0.6 mg in FCA is given similarly. This dose, dissolved in phosphosaline buffer, is then injected into the footpads. Subsequent booster injections (0.6 mg of conjugate) are administered in part into the rear thighs (intramuscularly), and in part into the upper-back region (subcutaneously). A second procedure involves a standard initial immunization as just described, followed by the injection of 0.6 mg of conjugate in FCA, partly into the rear footpads, partly into the rear thighs (intramuscularly), and partly into the upper-back region (subcutaneously). Some animals have been administered a third injection identical to the second, which had been given only three weeks after the initial treatment. In each of these two latter groups, good antibody titer has appeared earlier (i.e., as early as after the first booster injection) than in animals immunized according to the first scheme. Whether this situation is indicative of variability in the groups of rabbits or of significant differences in immunization plans is not known.

Goats were immunized by an initial injection of 0.1 mg of nucleotide-HSA conjugate in FCA into each of four surgically exposed lymph nodes in the posterior cervical and high inguinal regions. Six weeks later, a total of 1 mg of conjugate in FCA was administered subcutaneously in the regions of the initial incisions. Subsequent booster injections were given subcutaneously at 6-week intervals, each with a total of 1 mg of material in saline. Good antibody titers were obtained as early as after the second booster injection.

D. Isolation of Antibody

Blood is allowed to clot overnight at 4°C, and the serum is drawn off with a Pasteur pipet. Residual red and white blood cells are removed by centrifugation at 4°C at 100 x g for 5 min and 500 x g for 10 min, respectively. If the serum antibody titer is such that stock solutions diluted 1:2000 or more are usable, isolation of the globulin fraction is not necessary [13]. Otherwise, the presence of cyclic AMP phosphodiesterase activity in serum [3] becomes a consideration.

The globulin fraction is obtained by precipitation with 33% saturated ammonium sulfate [14]. Repeated precipitations cause losses in antibody protein. (We find that repeated precipitation of serum by ammonium sulfate reduces phosphodiesterase activity by a factor of 10. However, the initial (serum) phosphodiesterase activity is not high (1-5% of the specific activity of a 37,000 x g rat brain homogenate supernatant, for example), and a single precipitation appears sufficient to give reliable results.) The globulin precipitate is redissolved (and diluted) in 0.05 M sodium acetate buffer, pH 6.2. Dialysis of these solutions to remove residual ammonium sulfate does not appear to be necessary.

The globulin fractions are stored at -20°C.

E. Immunoassay Procedure

1. Standard Procedure

The "standard" procedure follows that reported by Steiner et al. [3]. A suitable dilution of rabbit cyclic AMP antibody (such that 30-60% of the intact marker radioactivity is bound) in 0.05 M sodium acetate buffer, pH 6.2 (0.1 ml), is added to 0.3 ml of the same pH 6.2 acetate buffer plus up to 0.1 ml of

[^{3}H]-cyclic AMP (0.5-1.0 pmole) or [^{125}I]-SCAMP-TME (0.05-0.1 pmole) diluted in acetate buffer. As little as 5-20 μl of marker can be added, but the tube-to-tube variability of added counts is probably decreased somewhat by the use of volumes closer to 100 μl. Unlabeled cyclic AMP (as a standard) or an extract of biological material may be present, if desired, in the 0.3 ml of buffer aliquot, or as an additional 5-20 μl volume. After the addition of antibody, the tubes are incubated for 2-4 hr at 4°C, and then 0.1 ml of the globulin fraction of goat anti-serum to rabbit IgG is added (in excess), and the incubation is continued overnight (16-18 hr total). (The IgG fraction of goat anti-serum to rabbit IgG may be purchased directly from Miles Laboratories, or the globulin fraction may be isolated from suitable goat serum by precipitation with ammonium sulfate (see Section II. D). Insufficient goat antibody titer should be suspected if the bound counts increase with additional aliquots of goat antibody or with dilutions of cyclic AMP antibody. Goat antibody would, of course, require the use of anti-serum to goat IgG.) The resulting cyclic AMP antibody-nucleotide-precipitating antibody complexes are sedimented by centrifugation at 2000 x g at 4°C for 30 min. The precipitates are washed once with 0.5 ml of acetate buffer and resedimented. If [^{125}I]-labeled marker was used, the precipitates may be counted directly in a gamma ray spectrometer. If [^{3}H]-labeled marker was used, the complexes are solubilized by the addition of 0.5 ml of 1.0 N NaOH, with heating to 70°C for 1 hr. The solutions are diluted with 15 ml of dioxane-based scintillation mixture [100 g of naphthalene, 14 g of 2,5-diphenyloxazole (PPO), and 0.1 g of 1,4-bis-[2-(4-methyl-5-phenyloxazolyl)]-benzene (dimethyl POPOP) per 2 liters of dioxane], and counted.

The incubation procedures may be shortened considerably unless maximal sensitivity is required [13]. In our laboratory, the antibody, nucleotide, and precipitating globulin have been incubated together routinely for a total of only 2-3 hr with reproducible and reliable results [15].

A standard immunoassay curve in which the number of counts bound by antibody is plotted against the number of pmoles of cyclic AMP added (on a semilogarithmic scale) is shown in Figure 1. A more detailed curve is available in the report by Steiner et al. [3], and shows a linear relationship between the cpm bound and the amount of unlabeled cyclic AMP added, over the range of 2-100 pmoles of added nucleotide.

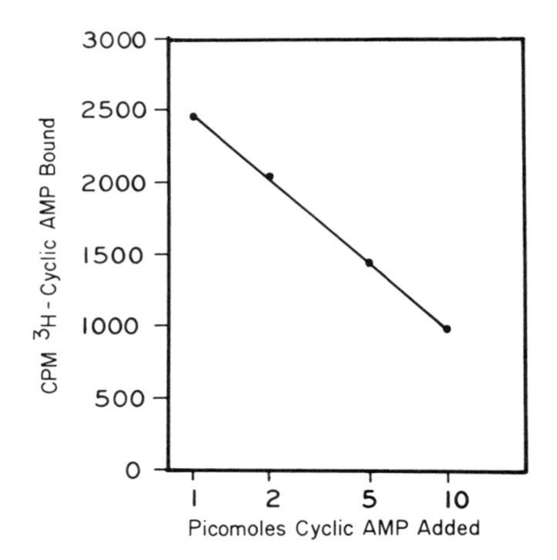

Fig. 1. Radioimmunoassay standard curve, showing the decreasing amounts of [^{3}H]-cyclic AMP bound to antibody in the presence of increasing amounts of unlabeled cyclic AMP. About 2970 cpm were bound in the absence of unlabeled nucleotide.

Linearity over the range 0.05-5 pmoles is also possible under conditions where high sensitivity is desirable and achievable [13]. The agreement between duplicate determinations is generally within 5%.

2. Millipore Filter Modification

We have recently determined [15] that the antibody-cyclic nucleotide complex may be retained on cellulose ester (Millipore) filters, by analogy with the protein kinase binding assay [4], without the need for precipitation by a second antibody. The convenience of this modification, along with the use of very short antibody-nucleotide incubation times (see following discussion) makes this procedure as rapid as the protein-binding procedure. Reproducibility is comparable to results obtained with the double-antibody radioimmunoassay technique. The Millipore filter method has also been adapted for use in the radioimmunoassay of prostaglandins [22].

The procedure described herein is suitable for the construction of standard curves and for the assay of cyclic AMP in samples of urine, but must be modified in some cases for the determination of cyclic AMP in tissue extracts (see Section II.G).

Radioactive (^{3}H or ^{125}I) marker is diluted with 0.05 M acetate buffer, pH 6.2, so that 8000-12,000 cpm (1 pmole or less) are present in 0.2 ml of buffer. To a 0.2-ml aliquot of buffer plus marker in a small test tube is added 5-10 μl of unlabeled cyclic AMP, standard or unknown, followed by 100 μl of an appropriate dilution of cyclic AMP antibody in acetate buffer (again, 30-60% of the intact marker radioactivity should be bound in the absence of added unlabeled cyclic nucleotide). The tubes are incubated for 1 hr at 4°C. (The binding of [^{3}H]-cyclic AMP to

the antibody after a 30-min incubation is more than 97
of that observed after 3 hr. An incubation period of
18 hr does not increase assay sensitivity beyond that
obtained at 1 hr. Incubation for 1 hr at room temperature appears to reduce somewhat the proportion of
marker cpm bound, relative to the same incubation at
0°C, but the assay sensitivity (i.e., the slope of
the curve in Figure 2) is not appreciably lowered.)
Then 1 ml of acetate buffer (pH 6.2) is added to each
assay tube, and the solutions are allowed to reequilibrate for 20 min at 4°C. Dilution causes a decrease
in the binding of [^{3}H]-cyclic AMP to the antibody of
5% within 1-5 min, and of an additional 3% within
5-15 min. From 15 to 60 min, binding does not change
further. Next, the contents of each tube are poured
onto a 25-mm (diam.) Millipore HAWP-025-00 (0.45 μm
pore size) filter (A sampling manifold available
from Millipore allows the filtration of up to 30
samples essentially simultaneously.), and the filter
is washed with 10 ml of the acetate buffer (usually
in four equal portions). The filters, to which the
antibody-cyclic AMP complexes are bound, may be
placed in thin butyl-plastic tubes (Lusteroid Container Co., Irvington, N. J.) and counted directly
in a gamma-ray spectrometer if [^{125}I]-labelled marker

is used. If $[^3H]$-labeled marker is used, the filters may be dissolved by addition to 1 ml of 2-ethoxyethanol in a scintillation vial. Solution of the filters in ethyl cellosolve is usually complete after 30 min. Others have also used 2-methoxyethanol for this purpose, and claim that it has superior solubilizing properties [7]. The dissolved filter is diluted with 15 ml of dioxane-based scintillation fluid (see Section II.E.1), and counted in a scintillation spectrometer. Occasionally, a filter either is defective or sustains damage during handling; the result is an anomalously low amount of radioactivity bound. Data from these filters should not, of course, be relied upon.

A convenient method of showing the data [16] is to plot the ratio of the cpm bound in the absence of added unlabeled cyclic nucleotide to the cpm bound in the presence of i pmoles of unlabeled cyclic AMP (B_o/B_i) against the quantity i pmoles of cyclic AMP added. Figure 2 illustrates this method, with data obtained from double-antibody (●) and Millipore filter (O) assays. An advantage of this graphical procedure is the linear plot, even in the 0-2 pmole region. In computing the (B_o/B_i), it is necessary to subtract assay blanks from B_o and B_i if they

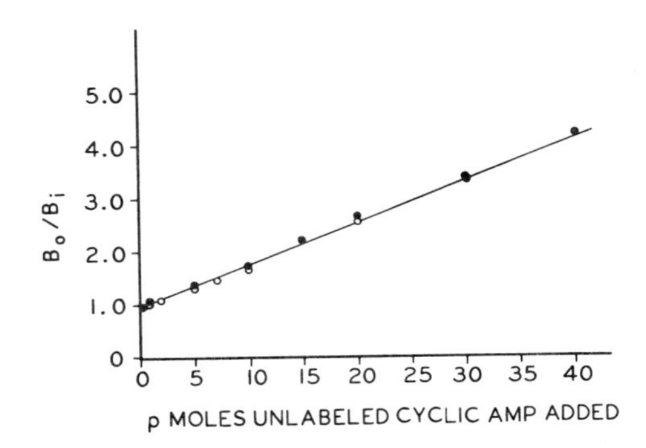

Fig. 2. Standard radioimmunoassay competition data for the precipitating antibody (●) and the Millipore filter (O) methods. The B_o/B_i is the ratio of net binding of $[^3H]$-cyclic AMP by antibody in the absence and presence of i pmoles of unlabeled cyclic AMP (taken from Ref. [15] with permission of Academic Press).

represent significant radioactivity relative to the lowest B_i value used; otherwise linearity will not be maintained at high (B_o/B_i) values. If $[^3H]$-cyclic AMP is used as the marker, a blank assay with the antibody absent will represent 1-2% of the B_o value. The use of normal IgG (i.e., from an unimmunized animal) in concentrations equal to that of the antibody aliquot gives blanks only slightly (if at all) higher than the "no-antibody" blank. It is often unnecessary to subtract these blank values from B_o and B_i in computing true (B_o/Bi) values. The situation with [^{125}I]-SCAMP-TME as marker is quite different. The "no-antibody" blank for this marker may be high, sometimes representing 20% or more of the B_o value. Markedly lower (although still significant) blank values are obtained by the addition of normal IgG or by the addition of very high amounts of unlabeled cyclic AMP (500 pmoles or more) to the standard assay system. Either of these approaches may give a more correct blank than the "no-antibody" value when the iodinated nucleotide derivative is used as marker. The slope of the resultant plot is a measure of the sensitivity of the assay, and will depend on the particular antibody preparation, the marker used, and, for the case of the [^{125}I]-SCAMP-TME marker, the integrity of the particular sample used. For the same sample of antibody, the slope of the plot will be two to three times as great with [^{125}I]marker as with [^{3}H]marker, and only about 1/100 of the sample antibody will be necessary with the iodinated nucleotide derivative.

Either standard curve is used in the same manner to determine the cyclic AMP content in an unknown sample. The addition of an aliquot of an unknown sample to an assay results in a certain (lower) number of cpm bound; the standard curve translates this binding into a number of picomoles of cyclic AMP added, and knowledge of the sample volume suffices to calculate the concentration of cyclic AMP or the amount in the total sample.

3. Other Modifications

It is possible to isolate the antibody-nucleotide complex by precipitation with ammonium sulfate [35]. This technique has been used in at least two other radioimmunoassay procedures for small molecules [17,18]. Precipitation is accomplished by the addition of normal rabbit IgG (carrier) followed by ammonium sulfate to 50% of saturation, for example, by the addition of a volume of saturated ammonium sulfate solution equal to that of the assay solution. The precipitate may be washed with a 50% saturated ammonium sulfate solution. The use of polyethylene glycol to precipitate the antibody-nucleotide complex has been described recently [25].

F. Antibody Specificity

The quality of a particular antibody is, in large measure, determined by its degree of specificity, i.e., the extent to which compounds similar in structure to cyclic AMP do (or do not) bind to the antibody.

At least two quantitative measures of the cross-reactivity of the antibody may be considered: (a) picomoles of compound necessary to lower binding by 50%, and (b) molar excess of compound (over marker) necessary to lower marker binding significantly (e.g., 5%). Steiner et al. [3] have used a measure similar to the inverse of (a) to determine the cross reactivity of various nucleotides and nucleosides against their antibody preparations. Thus, the relative binding affinity of ATP was 0.005%, of 5'-AMP was 0.0025%, and of cyclic GMP was 0.01%. The other naturally occurring nucleotides and nucleosides tested showed cross-reactivities of 0.01% or less.

In our laboratory, the specificity of a new antibody preparation is routinely checked against ATP, 5'-AMP, and cyclic GMP. Usable antibody should show a relative binding affinity of no greater than 0.005% (0.002% or less is desirable) for ATP. Suggested binding affinities for 5'-AMP and cyclic GMP are 0.01% or less. However, an antibody may possess a relative affinity for cyclic GMP much higher than 0.01% and still be usable. Since cyclic GMP occurs in tissues at concentrations lower than those of cyclic AMP, a relative affinity of about 1% should suffice in ordinary use. Very low affinities for cyclic GMP may be desirable in particular situations, such as the simultaneous radioimmunoassay of cyclic AMP and cyclic GMP concentrations in tissue or blood [13]. Our best antibody preparations have [in terms of measure (b)] shown less than a 10% reduction in marker bound in the presence of 5×10^5 and 5×10^4 molar ratios of ATP and cyclic GMP, respectively, and no significant decrease at all at molar ratios of 5'-AMP up to 3×10^4 [15].

G. Assay of Cyclic AMP in Biological Samples

1. Treatment of Biological Samples

How do we ensure that the cyclic AMP content of the biological samples of interest accurately reflects the cyclic AMP content present *in vivo* immediately before death of the animal and/or removal of the tissue or fluid? How do we ensure that the cyclic AMP content of the samples is accurately determined? The task of answering the first question has never been trivial and still presents significant difficulties for certain tissues; however, since this problem is relevant to all methods of assay for cyclic AMP, and not to the radioimmunoassay (or protein kinase binding assay) in particular, we shall not discuss it in detail. The interested reader is referred to the discussion by Broadus et al. [19] of the handling and storage of samples of extracellular fluids (e.g., blood and urine), to remarks dealing with the fixation of tissues _in situ_ [20,21], and to Chapter 1, section I.A for further comments on tissue preparation. We have examined the technique of rapid heating by microwave irradiation (Litton Menumaster Oven, Model 70/50) as a means of fixing rat brain tissues _in situ_ [20], and find it superior to immersion in either liquid nitrogen or isopentane cooled by liquid nitrogen, in terms of convenience and reproducibility of results.

The second question concerns methods of extraction of cyclic AMP from tissue, among other considerations. One method of extracting acid-stable, low-molecular-weight substances from tissue follows. Up to 50 mg of tissue is homogenized in 1 ml of cold (0°-4°C) 5% trichloroacetic acid (TCA) for about 30 sec. Insoluble macromolecules are sedimented by centrifugation at 2000 x g for 10 min, and the supernatant is drawn off and extracted repeatedly with 5-10 ml of ethyl ether saturated with water, until the pH of the aqueous phase approaches 6. (Repeated extraction of homogenates in TCA with anhydrous ethyl ether results in large losses in the weight of the aqueous phase. These losses are probably owing to the extraction of water (but not cyclic AMP) into the ether phase.) The aqueous phase is then heated to 70°-90°C for 2-10 min. An aliquot of this extracted aqueous phase (after cooling) may be used directly in the immunoassay. If the desired aliquot is not small (∿10% or less) compared to the standard assay volume, the sample may be evaporated to dryness with a stream of nitrogen and then dissolved in the appropriate amount of 0.05 M acetate buffer, pH 6.2, for use in the assay [1]. Concentration of the sample in this manner may, however, be difficult to accomplish quantitatively, and it is prudent to include a recovery marker such as [^{32}P]-cyclic AMP, either as a routine practice or to check technique initially. The amount of labeled cyclic AMP recovered can be calculated from the specific activity; if significant, it can be subtracted from the total cyclic AMP measured in the sample. Alternatively, if the concern with larger sample aliquots is primarily one of control of the pH of the incubation mixture, the addition of a small amount of concentrated acetate buffer of slightly higher pH (e.g., 6.4) to the extracted aqueous phase is recommended, so that the final buffer concentration and pH are 0.05 M and 6.2, respectively. (Steiner et al. [3] found that binding of ^{125}I-labeled marker to antibody was optimal over the pH range 5.5-6.2, and was reduced markedly at more alkaline pH values. We find that the addition of TCA to final pH values of 5.2 and 4.8 also lowers the binding of [^{3}H]-cyclic AMP to antibody, to 86% and 77%, respectively, of that observed at pH 6.2. The binding of [^{125}I] marker is similarly affected.)

Plasma samples may be extracted with an equal volume of 0.6 N perchloric acid; the extracts are applied to a column of Dowex 50 prepared as described in Section II.G.2, and the cyclic AMP fraction (fourth through eighth ml) is eluted with water and collected [13]. Extraction of plasma with TCA sometimes gives spurious results [13].

Urine may be added directly to the assay. Small volumes (1-10 μl) are necessary because the concentration of cyclic AMP in (human) urine is in the micromolar range [15]. Dilution of the urine samples with buffer may be desirable in some cases. In contrast to the high cyclic AMP concentrations in urine, the cyclic AMP concentration in various tissues of the rat (for example) is of the order of 1 pmole/per mg of tissue (wet weight) [3].

2. Assay of Samples: Problems and Solutions

If samples of rat cerebral cortex [22] or isolated lipocytes [23] that have been homogenized in 5% TCA are studied with the radioimmunoassay technique, it is found that the addition of a sample aliquot *raises* the B_o value, instead of lowering it owing to the presence of cyclic AMP in the sample. The cpm bound to antibody in the absence of the sample are typically raised by 10-20% in the presence of a sample aliquot. This phenomenon is not abolished by homogenization in up to 12% TCA, or in 0.4 N perchloric acid, nor is prior filtration through a 0.45- μm Millipore filter successful. The addition of 0.13 M EDTA to the assay was also unsuccessful. (The EDTA caused a 30% decrease in the binding of marker to antibody recovered in the absence of sample, but did not prevent an enhancement of this reduced binding by the sample aliquot.) An "antibody-binding-enhancement factor" appears to be present in the sample; it is not precipitated by up to 12% TCA nor destroyed by heating to 90°C after extraction with ether.

It was found, first for extracts of lipocytes [23] and then for extracts of cerebral cortex [22], that this factor could be removed by passage of the supernatants from the homogenizations in TCA through a 8-mm inner diameter column of Dowex 50W-X8 (100-200 mesh), prepared with 4 ml of a 50% slurry of resin in water, prior to assay. The cyclic AMP could be eluted with water. (Pilot experiments with TCA containing [^{3}H]-cyclic AMP, and with or without tissue extracts, indicated that up to 80% of the cyclic AMP was eluted with the fifth through eighth ml of water. Development of the column with up to 12 ml of 0.05 M sodium acetate buffer, pH 6.2, did not elute [^{3}H]-cyclic AMP in substantial quantities. It may be necessary, depending on the characteristics of the tissue sample with respect to the factor and the concentration of homogenate, to use longer Dowex-50 columns in some cases. In such an event, the elution pattern of [^{3}H]-cyclic AMP should be determined under the particular conditions.) This step also removed the TCA and, hence, made ether extraction unnecessary. Treatment of the sample with Dowex-50 also effects a major purification of cyclic AMP relative to the other common nucleotides, and, accordingly, batches of antibody previously considered undesirable because of poor nucleotide specificity may become usable.

In a routine assay procedure in which [^{3}H]-cyclic AMP is used as binding marker, about 10,000 cpm of [^{32}P]-cyclic AMP per ml of 5% TCA is used as a recovery standard. (If [^{125}I]-SCAMP-TME is used as the binding marker, and if a gamma-ray spectrometer is available, then it is possible to use [^{3}H]-cyclic AMP as the recovery standard. Its advantages over [^{32}P]-cyclic AMP of higher specific activity and much longer half-life are partially counterbalanced by the need to consider counting efficiencies in correcting the raw data.)

Up to 50 mg of tissue is homogenized in 1 ml of TCA containing the recovery standard, and 0.5 ml is applied to the Dowex-50 column. The column is developed with distilled water, and the first 4.5 ml eluted are discarded. The next 3 ml are collected and 1 ml of 0.2 M sodium acetate buffer, pH 6.4, is added to this fraction, yielding a final buffer concentration of 0.05 M and a pH of 6.2. Of this, 1 ml is counted for recovery. To a 2-ml aliquot is added 50 μl of antibody, and the assay is then conducted as described in Section II.E.2, except that a 3-hr incubation is used. This time is not necessarily optimal, and represents a compromise between a conveniently short period and one long enough to closely approach equilibrium binding. The use of larger volumes also eliminates the need to add additional buffer to the assay tube to ensure quantitative transfer to the filter.

The nature of the antibody-binding-enhancement factor is obscure. Extracts of rat cerebral cortex also enhance the binding of [^{3}H]-cyclic AMP to bovine cerebral-cortex protein kinase [22]. This result, together with the heat stability of the factor, suggests a resemblance to the protein inhibitor of cyclic AMP-dependent protein kinases reported by Walsh et al. [24]. This protein inhibits kinase activity, but enhances binding of cyclic AMP to the enzyme complex. The factor discussed here, however, differs from the protein kinase inhibitor in its solubility in 12% TCA.

It may be possible, by analogy with the use of a partially purified kinase inhibitor preparation in the protein kinase binding assay, as suggested by Gilman [4], to isolate a crude antibody-binding-enhancement factor fraction and add it to the assay system described in Section II.E.2. If this addition results in maximized marker-antibody binding, extracts prepared as discussed in Section II.G.1 may then be directly usable.

III. CYCLIC AMP-DEPENDENT PROTEIN KINASE BINDING ASSAY

The simplicity of the protein kinase binding assay for cyclic AMP described by Gilman [4], together with its high sensitivity, has led to its rapid acceptance. The overall specificity of this procedure, although lower than that of the radioimmunoassay, is probably adequate for most applications (see Section III.D).

In principle, the protein kinase from any tissue may be used. Skeletal muscle is an advantageous choice because of its favorable binding constant for cyclic AMP [4]. In our laboratories, we use a kinase preparation from bovine cerebral cortex that appears to work well, although the binding constant for cyclic AMP (measured by stimulation of kinase activity) is higher, by a factor of about 3, than that for skeletal muscle [29]. Other mammalian tissues have been used as sources of kinase [5], and many appear to be usable [28]. A protein kinase binding assay for cyclic GMP, which uses the enzyme from lobster muscle, has also been described [7].

The assay procedure with the kinase as binding protein uses the same rationale as the radioimmunoassay.

One difference lies in the addition of a heat-stable protein fraction ("protein inhibitor fraction"), which improves the binding of [^{3}H]-cyclic AMP to the kinase by a factor of 2 [4], although it inhibits the phosphorylation of acceptor proteins catalyzed by the kinase [24] (see also Section II.G.2).

A kit for measuring cyclic AMP by the method of Gilman [4] with (kinase) binding protein from bovine (skeletal) muscle is available commercially from Affitron Corporation (Costa Mesa, Cal.). The sales brochure suggests that the protein inhibitor fraction is not included in the kit. A similar kit for the assay of cyclic GMP is available from Diagnostic Products Corporation (Los Angeles, Cal.).

A. Preparation of Protein Kinase

Cyclic AMP-dependent protein kinase is partially purified from bovine cerebral cortex by the procedure of Kuo et al. [29] (see also Chapter 7). This methodology is applicable to the preparation of kinase from other tissues, including skeletal muscle. Frozen calf brain maintained at -70°C has been purchased from Miles Laboratories; about 150 g of cerebral cortex may be conviently worked up. The tissue is thawed, cut up, homogenized (in 3 ml/g of 4 mM EDTA, pH 7.0, for 2 min) in a Waring blendor, and centrifuged at 4°C for 20 min at 27,000 x g. The supernatant fraction is treated with 1 M acetic acid (with stirring) until the pH is 4.8, then allowed to stand for 10 min. The precipitate formed is sedimented by centrifugation at 4°C for 30 min (27,000 x g). The pH of the supernatant fraction is adjusted to 6.5 by the addition of 1 M potassium phosphate, pH 7.2. The volume at this point is about 350-400 ml. At this point, kinase activity is precipitated by the addition of solid ammonium sulfate (32.5 g/100 ml). The suspension is stirred for 30 min, then centrifuged for 20 min at 4°C (27,000 x g). The pellet is dissolved in 6% of the crude extract volume (before precipitation) of 5 mM potassium phosphate buffer, pH 7.0, containing 2 mM EDTA. This solution is then dialyzed against the same buffer (20 vol) for 14 hr with two changes. Another centrifugation (27,000 x g, 30 min, 4°C) yields a supernatant fraction, which is analyzed for protein [34] before application to a column of DEAE-cellulose (we use Whatman DE-52) ion exchange resin (about 50 g resin/g protein), that has been equilibrated with the same buffer used for dialysis. The resin is equilibrated with four portions of buffer for 10 min, and then with one portion for 30 min, before a 5% (w/v) slurry of resin in buffer is poured into a column. After application of the enzyme solution, the column is washed with 2 bed-volumes of 0.1 M potassium phosphate, pH 7.0, containing 2 mM EDTA, and 5-ml fractions are collected. Elution is then continued with 0.3 M potassium phosphate, pH 7.0, containing 2 mM EDTA. Fractions were tested for their ability to bind [^{3}H]-cyclic AMP. Almost all of the binding activity is eluted by the 0.3 M buffer; the fractions with the high activity elute starting at 90-100 ml. These fractions are pooled and dialyzed overnight against two changes of 5mM potassium phosphate buffer, pH 7.0, containing 2 mM EDTA. This preparation binds about 34 pmoles of cyclic AMP per mg of protein. The enzyme was stored in 1-ml aliquots at -15°C.

B. Preparation of Protein Kinase Inhibitor

The isolation of a crude inhibitor fraction from bovine muscle [4] is described here; a factor inhibitory to kinase activity has also been reported for brain [31] and is present in several other mammalian tissues [29].

Bovine muscle is homogenized in 0.01 M tris buffer, pH 7.5, then boiled for 10 min. Particulate matter is removed by filtration or centrifugation, and the inhibitory activity is precipitated with 50% TCA (final concentration of 5% TCA). After sedimentation at 15,000 x g, the pellet is dissolved in H_2O, and the pH of the solution is adjusted to 7 with 1 N NaOH. The sample is dialyzed against distilled water at room temperature, and any precipitate is removed. This dialyzed fraction is used in the assay.

C. Assay Procedure [4]

High-specific-activity (14-24 Ci/mmole) [^{3}H]-cyclic AMP is diluted with 0.05 M sodium acetate buffer, pH 4.0, so that 1.0 pmole (10,000-15,000 cpm at 30-40% counting efficiency) or less is present in a 0.2-ml aliquot of buffer. To such a 200-μl aliquot in a disposable culture tube is added 50 μl of protein kinase preparation, sufficient to bind 20-50% of the counts present. It is convenient to add up to 50 μl of standard cyclic AMP solution or of tissue extract (prepared as outlined in Section II.G.1, except that the final extract should contain acetate buffer near pH 4.0). Protein kinase inhibitor fraction may also be present, if desired. The assay mixture, initiated by the addition of the protein kinase, is incubated at 0°C for about 1 hr. (The rapidity of approach to equilibrium will depend, in general, on such parameters as the particular kinase used, the assay volume, the amount of substrate (cyclic AMP), and the presence or absence of kinase inhibitor. In the absence of kinase inhibitor, a slow decrease in binding at times longer than 60 min may be observed with kinase preparations from bovine muscle [4] and cerebral cortex [22]. An incubation time of 30-60 min appears optimal for assay with the cerebral cortex kinase. In the presence of inhibitor, binding to bovine skeletal muscle kinase is stabilized at times longer than 60 min [4].) The tubes are then diluted with 1 ml of 0.02 M potassium phosphate buffer, pH 6.0. After 5 min, the contents of the tubes are poured onto Millipore filters (see Section II.E.2) previously rinsed with phosphate buffer. (Dilution causes a decrease in binding of the marker to the kinase from cerebral cortex that reaches about 25% 15 min after dilution. No further decrease results up to 60 min after dilution. From 3 to 10 min after dilution, binding decreases only about 8%, and a waiting time within this period may be used when higher sensitivity is desire; however, closer adherence to a consistent waiting time for all samples is necessary.) The filters are washed with 10-15 ml of buffer and allowed to dissolve in 1 ml of 2-ethoxyethanol (Cellosolve) in a scintillation vial for 30 min. According to Gilman [4], results are independent of the volume of buffer rinse over the range 3-20 ml. Dioxane-based scintillation fluid (see Section II.E for composition) is added (15 ml) and the samples are counted.

A typical standard curve obtained with the preparation from bovine cerebral cortex is illustrated in Figure 3. About 1 pmole of [^{3}H]-cyclic AMP has

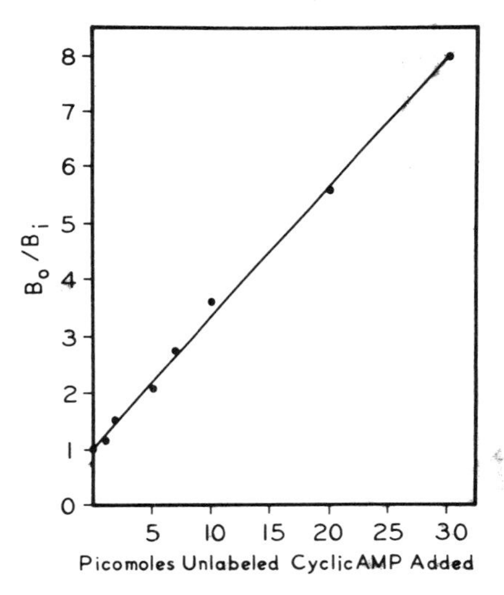

Fig. 3. Standard curve for protein kinase binding assay with the preparation from bovine cerebral cortex. Each tube contains about 1 pmole of [^{3}H]-cyclic AMP marker.

been used as a marker. The data are plotted, as in Figure 2, according to Hales and Randle [16]. The sensitivity with this assay under these conditions is comparable to that obtained with the radioimmunoassay with $[^{125}I]$-SCAMP-TME as marker. The addition of 1 pmole of unlabeled cyclic AMP to the marker lowers the counts bound in the absence of unlabeled nucleotide (B_o) by as much as 30%. Reduction of the amount of $[^{3}H]$-cyclic AMP (to amounts that are still saturating) and/or the assay volume increases the sensitivity of this procedure [4].

D. Kinase Specificity

The specificity of the bovine muscle kinase, as reported by Gilman [4], indicates that interference by 3',5'-cyclic nucleotides may be higher when kinases, rather than antibodies, are the binding proteins. For example, cyclic IMP and cyclic GMP inhibit binding of $[^{3}H]$-cyclic AMP to kinase by 50% at 8-fold and 125-fold molar excesses, respectively. In contrast, 50,000-fold molar excesses of cyclic GMP reduce binding to antibody by less than 10% (Section II.F). A similar situation appears to exist with respect to the cyclic pyrimidine nucleotides. In terms of the cross reactivities of the nucleotides used to characterize the specificity of antibody lots (ATP, 5'-AMP, and cyclic GMP), the use of the kinase as binding protein does not present serious disadvantages. The cross reactivity of 5'-AMP is very low, and that of ATP (50% inhibition of binding at a 25,000-fold molar excess) is probably adequate in most cases; dilution of the tissue extract by 10- to 50-fold, an option that is possible with the high sensitivity of the procedure [4], may be used to lower the ATP concentrations. The high cross reactivi- of cyclic GMP would appear, nevertheless, low enough for use in ordinary assay situations, in view of the lower concentrations of cyclic GMP in tissues relative to those of cyclic AMP.

E. Assay of Biological Samples

Much of Section II.G, on the assay of cyclic AMP in biological materials, may be profitably utilized in the discussion of procedures and possible pitfalls in cyclic AMP assays involving kinase binding proteins. The finding that extracts of rat cerebral cortex from homogenization of tissue in 12% TCA enhanced the binding of [^{3}H]-cyclic AMP to bovine cerebral cortex kinase [22], suggests that substances other than the kinase inhibitor (which precipitates in 5% TCA), that act like the inhibitor fraction in the assay, may be present in TCA extracts of tissue. Although our experience in the assay of cyclic AMP levels in biological samples with the kinase binding proteins has been limited, and did not include use of the kinase inhibitor fraction, the possible occurence of other "binding enhancement factors" in tissue extracts prompts the recommendation that the kinase inhibitor fraction be added routinely to each assay tube, in amounts that are sufficient to maximally increase the binding of [^{3}H]-cyclic AMP to kinase. The improved performance of the assay with inhibitor present that has been noted [4] may arise, in part, from the ability of the inhibitor to prevent further enhancement of binding by other tissue factors.

Excessive salt concentrations can interfere with the binding of cyclic AMP to kinase, so that tissue extracts, for example, should be concentrated with caution prior to addition to the assay [32]. This problem may be circumvented by passage of the sample through a 0.4 x 3 cm column of Dowex-50 ion-exchange resin [32]. Samples will, of course, be desalted if interfering substances in the tissue extracts necessitate chromatography on Dowex-50 before assay, as outlined in Section II.G.

ACKNOWLEDGMENT

What success we have enjoyed in adapting these procedures for use in our laboratory is owing to large measure to the time and efforts of many of the personnel of the Squibb Institute for Medical Research, as well as of several colleagues outside the Institute. The continued interest of Dr. S. M. Hess has been particularly helpful in the completion of this project.

NOTES

[1]Catalog No. S-37148-3, Alfred Bader Chemicals (division of Aldrich Chemical Co.) or from Sigma Chemical Co. Attempts to prepare MDCC from morpholine and dicyclohexylcarbodiimide by refluxing in t-butyl alcohol-water solvent following general guidelines [8,9] gave equivocal results. MDCC solubilizes cyclic AMP in pyridine; triethylamine is also reported to serve this purpose [1,10]. Prolonged heating (about 2 hr) is necessary for dissolution. Cyclic GMP may be partially solubilized with trioctylamine [1].

[2]Chilling of the reaction mixture in ice before addition of the water has also been recommended. Addition of the water destroys excess succinic anhydride, plus any carboxylic acid-phosphate mixed anhydride [10].

[3]Recovery of 2'*O*-succinyl cyclic AMP from the reaction mixture by ion-exchange chromatography on (Pharmacia) Sephadex QAE-25 has been described [11]. Also, this product can be crystallized from aqueous solutions during concentration before any lyophilization [30].

[4]A possible side reaction leads to the formation of N^6,2'*O*-disuccinyl cyclic AMP, particularly if the acylation time is prolonged [10]. Such a product is distinguishable from the 2'*O*-succinyl cyclic AMP by its relative resistance to complete hydrolysis by base; treatment of 2'*O*-succinyl cyclic AMP with 0.1 N NaOH causes complete reversion to cyclic AMP [3,10]. In addition, N^6 substitution causes the region of maximal UV absorption to shift from 258-260 nm to 270-273 nm [10].

[5]The highest specific activity consistent with reasonable radiochemical stability is desirable in a marker, since a measure of the sensitivity of the assay is the lowest quantity of unlabeled nucleotide able to significantly decrease binding of the marker by the antibody. The greater the radioactivity per picomole of marker, the smaller the number of molecules necessary to give the desired level of bound radioactivity, and the smaller the number of unlabeled molecules necessary to compete successfully for antibody binding sites.

[6]Use of a purified sample of tyrosine methyl ester is recommended. A commercial sample of the hydrochloride salt in solution, when treated with bicarbonate, liberates the tyrosine methyl ester, which precipitates. Any tyrosine present remains soluble. The insoluble tyrosine methyl ester may be recrystallized from benzene and is fairly stable in the dry state [30]

REFERENCES

[1] A. L. Steiner, C. W. Parker, and D. M. Kipnis, Advan. Biochem. Psychopharmacol., 3, 89 (1970).

[2] G. A. Robison, R. W. Butcher, and E. W. Sutherland, Cyclic AMP, Academic press, New York, 1971, pp. 456-478

[3] A. L. Steiner, D. M. Kipnis, R. Utiger, and C. Parker, Proc. Natl. Acad. Sci. U. S., 64, 367 (1969)

[4] A. G. Gilman, Proc. Natl. Acad. Sci. U. S., 67, 305 (1970)

[5] G. M. Walton and L. D. Garren, Biochemistry, 9, 4223 (1970)

[6] B. L. Brown, R. P. Elkins, and W. Tampion., Biochem. J., 120, 8P (1970)

[7] F. Murad, V. Manganiello, and M. Vaughan, Proc. Natl. Acad. Sci. U. S. , 68, 736 (1971)

[8] J. G. Moffatt and H. G. Khorana, J. Am. Chem. Soc., 83, 649 (1961)

[9] J. G. Moffatt and H.G. Khorana, J. Am. Chem. Soc., 83, 663 (1961).

[10] J.-G. Falbriard, Th. Posternak, and E. W. Sutherland, Biochim. Biophys. Acta, 148, 99 (1967)

[11] H. Cailla and M. Delaage, Abstract, Int. Conf. Physiology and Pharmacology of Cyclic AMP, Milan, Italy, 1971.

[12] G. D. Fasman, Biol. Macromol., 1, 499 (1967).

[13] A. L. Steiner, personal communication.

[14] H. F. Deutsch, in Methods in Immunology and Immunochemistry (C. A. Williams and M. W. Chase, eds.), Academic Press, New York, 1967, p. 319.

[15] I. Weinryb, I. M. Michel, and S. M. Hess, Anal. Biochem., 45, 659 (1972).

[16] C. N. Hales and P. J. Randle, Biochem. J., 88, 137 (1963).

[17] S. Spector and C. W. Parker, Science, 168, 1347 (1970).

[18] B. M. Jaffe, J. W. Smith, W. T. Newton, and C. W. Parker, Science, 171, 494 (1971).

[19] A. E. Broadus, J. G. Hardman, N. I. Kaminsky, J. H. Ball, E. W. Sutherland, and G. W. Liddle, Ann. N. Y. Acad. Sci., 185, 50 (1971).

[20] M. J. Schmidt, D. E. Schmidt, and G. A. Robison, Science, 173, 1142 (1971).

[21] G. A. Robison, R. W. Butcher, and E. W. Sutherland, Cyclic AMP, Academic Press, New York, 1971, p. 461.

[22] I. Weinryb and I. M. Michel, unpublished experiments.

[23] M. Chasin, personal communication.

[24] D. A. Walsh, C. D. Ashby, C. Gonzalez, D. Calkins, E. H. Fischer, and E. G. Krebs, J. Biol. Chem., 246, 1977 (1971).

[25] B. Desbuquois and G. D. Aurbach, J. Clin. Endocrinol. Metab., 33, 732 (1971).

[26] R. Szczesniak, personal communication.

[27] W. M. Hunter and F. C. Greenwood, Nature, 194, 495 (1962).

[28] L. D. Garren, Ann. N. Y. Acad. Sci., 185, 69 (1971).

[29] J. F. Kuo, B. K. Krueger, J. R. Sanes, and P. Greengard, Biochim. Biophys. Acta, 212, 79 (1970).

[30] J. T. Sheehan, personal communication.

[31] E. Miyamoto, J. F. Kuo, and P. Greengard, J. Biol. Chem., 244, 6395 (1969).

[32] A. G. Gilman, personal communication.

[33] J. Fried, personal communication.

[34] O. H. Lowry, N. J. Rosebrough, A. L. Farr, and R. J. Randall, J. Biol. Chem., 193, 265 (1951).

[35] A. L. Steiner, C. W. Parker, and D. M. Kipuis, *J. Biol. Chem.* **247**, 1106 (1972).

Chapter 3

HIGH-PRESSURE LIQUID CHROMATOGRAPHY FOR THE ANALYSIS OF CYCLIC NUCLEOTIDES

Gary Brooker

Department of Pharmacology
University of Virginia
School of Medicine
Charlottesville, Virginia 22901

I. INTRODUCTION

High-pressure chromatography is most useful in investigations involving cyclic nucleotides. It allows for the direct quantitation of these nucleotides, based upon their ability to be eluted from an anion-exchange column with detection of their inherent ultraviolet absorptivity by an extremely sensitve, UV-absorption, flow-cell detector. This chromatographic system makes it possible to easily measure picomole quantities of cyclic nucleotides, other nucleotides, and cyclic-nucleotide analogs.

The basic high-pressure anion-exchange chromatography system used in my laboratory consisted of a Varian Aerograph LCS 1000 Liquid Chromatograph with an improved UV flow-cell detector, a Varian Aerograph 10-in Potentiometric Recorder, and a Hewlett Packard 3370B Digital Integrator. This equipment is relatively trouble free; however, it is important to have a general knowledge of the system design and operation to preclude many service calls. In general, the ability to obtain good results at the picomole level is dependent upon optimum performance of the mechanical and electronic systems.

II. DESCRIPTION OF THE ANALYTICAL SYSTEM

High-pressure anion-exchange chromatography was performed with a 3-meter capillary "pellicular" anion-exchange column, which was originally described by Horvath, et al. [1]. In practice, a Varian Aerograph LCS 1000 Liquid Chromatograph with an 8-μl, 1 cm pathlength and 254-nm UV flow cell was used. The detector has been modified as previously described [2], and the resin in the precolumn removed to allow injection of 30-μl samples. It has been shown [3] that peak height measurements are directly proportional to the amount of cyclic (AMP) injected into the chromatograph. Generally, full-scale deflection of the recorder was set to equal 2×10^{-3} absorbancy units. Under optimal conditions the noise level is generally 2×10^{-5} absorbancy units when all systems are operating properly. The resin in the 3-meter column was converted to the chloride form by washing the column with 50 ml of 1 N HCl, and then to neutrality with 100 ml of distilled water. The complete LCS 1000 is probably not necessary for the analysis of cyclic AMP; the combination of many of the components found within the unit can be used to put together a less costly system capable of measuring cyclic nucleotides. Figure 1 shows the general schematic of the system, consisting of a solvent reservoir,

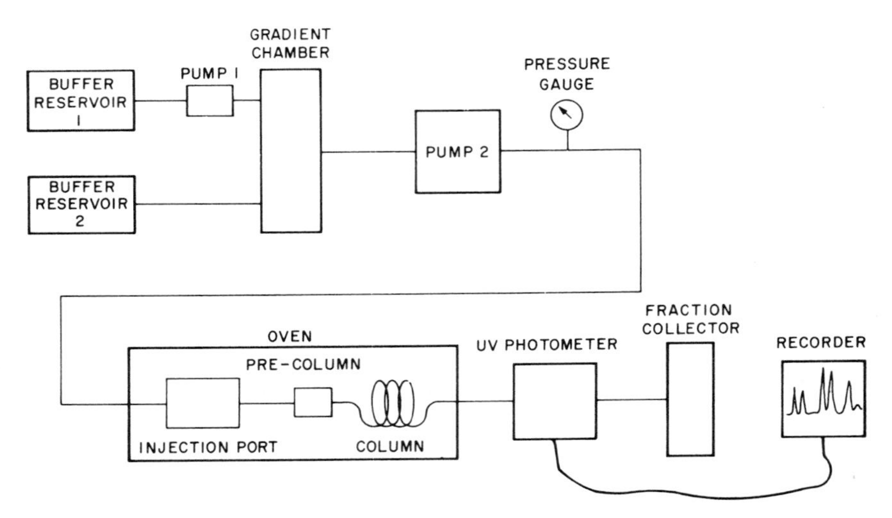

Fig. 1. Schematic diagram of high-pressure chromatography system (from *Varian Aerograph Nucleic Acid Constituents by Liquid Chromatography*, 1970, p. 3).

degassing chamber, high-pressure pump, injection port, "pellicular" ion-exchange column and oven, jacketed UV-absorption flow cell, fraction collector, digital integrator, and potentiometric recorder.

III. PRECAUTIONS

It is essential that the temperature of the high-pressure anion-exchange column be kept constant, since flow through the column will be altered by the resistance of the column itself. The resistance of the column decreases as the temperature is increased above ambient. I have found optimal operation of the "pellicular" anion-exchange column to occur at 80°C. It is also essential for optimal operation that no air bubbles or leakage from any fitting occur. We have found that operating the system for 0.5-1 hr at 1-3 times the pressure used for our analytical work tends to purge the system of any air bubble that might have developed and, in addition, allows us to check all fittings for any obvious leakage that might not be detected at the lower pressures. Generally, we operate the system at 24-25 ml/hr, a flow rate that generates a back pressure of about 1000 psi. It is easy to tell if air bubbles are present within the system, since it then does not come up to maximum pressure instantaneously when the pump is started; the base line appears rather unstable and a long delay occurs when the pump is turned off and the system begins to approach zero pressure. Increase of the pumping rate to 2500-3000 psi for 0.5 hr usually eliminates small air bubbles; however, if this treatment does not correct the situation more drastic purging of the system is necessary. Air bubbles seem to become easily trapped within lines in which active flow is not occurring, for example, the tube that leads to the pressure gauge. I have found that purging the system with carbon dioxide is an effective way to eliminate air trapped within these lines. A source of CO_2 is introduced into the lines in which the air bubbles are thought to reside. It is important not to purge the column itself. Once the system is reconnected and the solvent is pumped for 1-2 hr at very high pressure, the carbon dioxide then dissolves within the solvent and leaves the system free of air bubbles. All buffers should be degassed before introduction into the chromatograph, and the UV flow cell should be routinely checked by looking down into the cell to be sure no dirt or dust particles are found within the cell itself.

CAUTION: BE SURE TO PROTECT YOUR EYES FROM DANGEROUS UV RADIATION EMITTED FROM THE CELL!

Recently Varian Aerograph have designed the cell for more trouble-free operation. In the past, the gaskets separating the reference and sample cells eventually leaked and liquid from the sample cell invaded the reference cell. This, of course, caused unstable base-line conditions, since this instrument operates on a dual-beam principle. The UV source is a germicidal lamp obtainable either from Varian or from General Electric Company, with the absorption of the cell being detected by a dual-element photo resistor, as diagramatically shown in Figure 2.

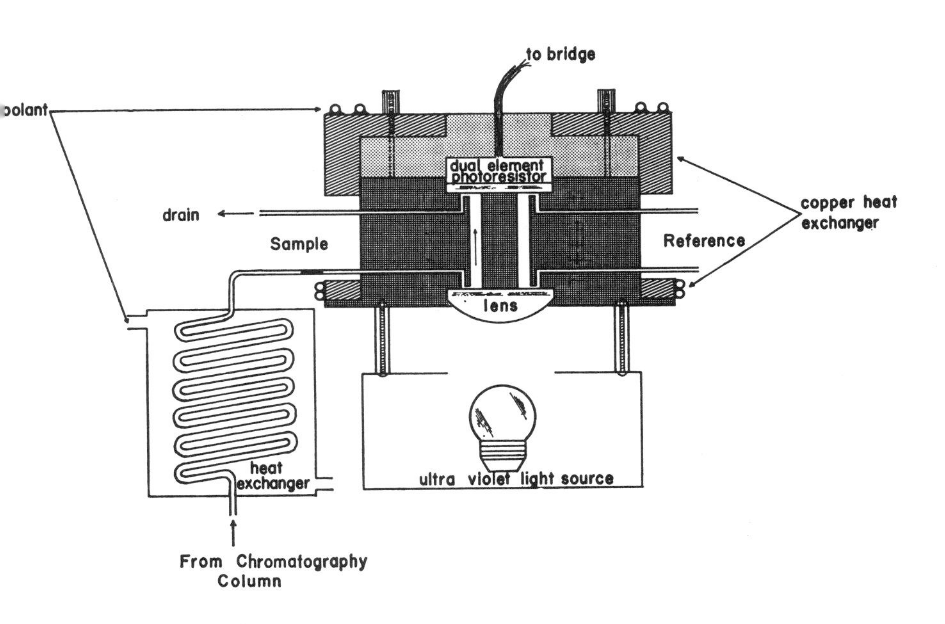

Fig. 2. Diagram of ultraviolet flow-cell detector (from Varian Aerograph UV Detector Instruction Manual, Pub. No. 85-001117-10, March 1972 p.40).

The dual-element photo resistors are uniquely sensitive to changes in temperature [2]. A jacket with a circulating water supply to maintain the cell and the sample at a constant temperature has increased the stability of the unit. During operation of the system the photo resistors should have a low resistance. Optimal sensitivity occurs when resistance across pins 1 and 2 or 2 and 3 are equal and between 20,000 and 30,000 ohms. However, more noise is seen when these resistances are 80,000 to 200,000 ohms. In addition, it may be that the photo resistors age over a prolonged time and their sensitivity to light decreases. If the system lacks the sensitivity and stability that had been previously demonstrated, then one might suspect a leaky cell, air bubbles, or an inherent electrical problem. It is conceivable that an unstable UV source could contribute to the noise level, even though most noises should be eliminated by the dual-beam operation. It generally takes one to two days of constant operation for a new lamp to stabilize; in addition, it is advisable never to turn off the UV source, so that stability is maintained from day to day.

IV. MEASUREMENT OF CYCLIC AMP

Each chromatographer will find that a slightly different concentration of HCl might be necessary for optimum elution of cyclic AMP from his particular batch of anion-exchange resin. In our system, HCl, pH 2.20, elutes cyclic AMP from this pellicular ion-exchange resin and retards cyclic AMP by about 2 to 3 min from the injection breakthrough. As shown in Figure 3, 43 pmoles of cyclic AMP give nearly a full-scale peak on the recorder; the response is linear from 0 to 100 pmoles of cyclic AMP, which can be quantitated simply by measuring the peak height or area of the chromatogram [4]. For the analysis of cyclic GMP, 0.2 M NaCl was added to the HCl solution. Under these conditions cyclic AMP elutes with the breakthrough point, as shown in Figure 4.

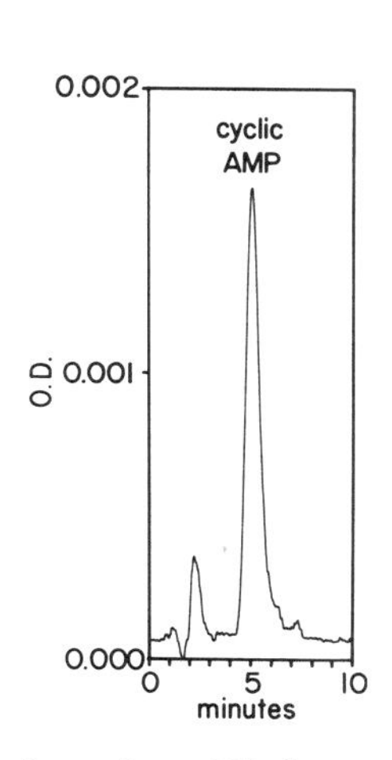

Fig. 3. High pressure chromatography of 43 pmoles of cyclic AMP (from ref. [2]).

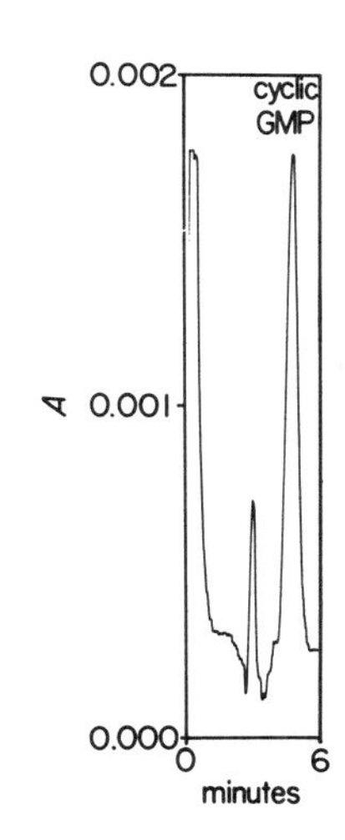

Fig. 4. High pressure chromatography of 50 pmoles of cyclic GMP (from ref. [7]).

Analogs of cyclic AMP can also be chromatographically separated with this system. Figure 5 shows that 8-io cyclic AMP [5] has been separated from 8-chlorocyclic

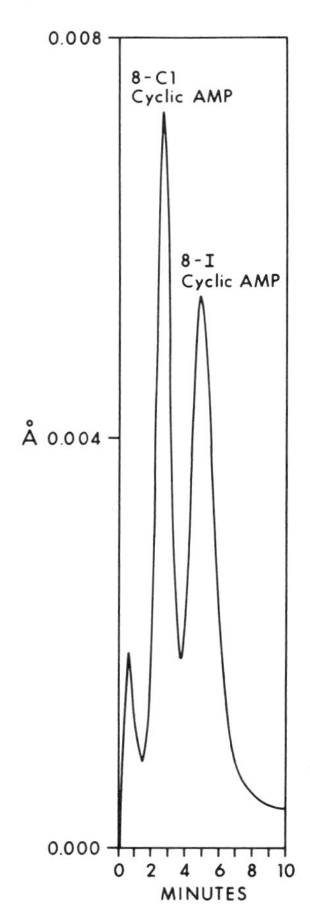

Fig. 5. High-pressure chromatographic separatio of 116 pmoles of 8-chlorocyclic AMP from 148 pmoles o 8-iodocyclic AMP.

A. Measurement of Cyclic AMP in Tissues and in Urine

We place 10-400 mg of frozen tissue in a 17 x 100 mm polypropylene test tube precooled in liquid nitrogen, add 10 μl of [^{3}H]-cyclic AMP tracer (0.1-1 pmole), and place the tube on the polytron homogenizer. We add 3 ml of 5% trichloroacetic acid, and the sample is then quickly homogenized at full speed for 10 sec. The homogenate is centrifuged in a benchtop centrifuge, and the supernatant is decanted to another identical tube. The precipitated protein is then dissolved in exactly 2 ml of 0.2 M NaOH, and digested for 1-12 hr. An aliquot of this solution is then used to determine protein by the method of Lowry et al. [6]. The trichloroacetic acid is extracted with three 8-ml portions of water-saturated diethyl ether. After each addition of ether, the tube is mixed on a variwhirl vortex mixer (Van Waters & Roger No. 5810-006) for 10 sec. After the ether layer separates, it is removed with a teflon-tipped suction aspirator. Then 200 μl of 0.5 M tris-HCl, pH 8, is added and the sample is applied to an anion-exchange column filled with AG 1 X 2, 50-100 mesh, chloride-form resin prepared in a disposable 0.8 x 5 cm pasteur pipet. Addition of tris-HCl to this solution brings the pH of the solution up above 6 or 7 so that cyclic AMP, when applied to the column, is not eluted by the low pH of the extract. A small propylene funnel (Arthur H. Thomas Co., No. 5587G) is attached to the top of the column and acts as a solution reservoir above the column. The solution is allowed to pass through the column, and 5 ml of water is applied. After the water, 10 ml of HCl, pH 1.3, is applied and collected to elute cyclic AMP from the column and to separate it from ATP in the tissue extract, which remains on the column. It is necessary to eliminate ATP from the sample before treatment with zinc-barium, to prevent the formation of nonenzymatically produced cyclic AMP from ATP [7]. To the cyclic AMP fraction, 100 µl of 2 M tris is added, followed by 2.5 ml of 5% zinc sulfate and 2.5 ml of 0.3 N $Ba(OH)_2$. It is important that the addition of the zinc and the barium exactly neutralize each other to the phenolphthalein end point. (However, do not add phenolphthalein to the samples that are being assayed. The precipitate is mixed and then centrifuged in a benchtop centrifuge. Water (10 ml) is applied to the column to ready it for reapplication of the $ZnSO_4$-$Ba(OH)_2$ supernatant. A new column is used when tissue samples have been prelabeled with [^{14}C]-

adenosine to determine the specific activity of [^{14}C]-cyclic AMP. The zinc-barium supernatant is applied to the column, followed by 30 ml of HCl, pH 2.7. Finally, 25 ml of HCl, pH 2.1 is applied to the column and collected into a 40-ml conical centrifuge tube (pyrex No. 8124). The flow rate for these columns is about 3 ml per min. Cyclic AMP fractions are evaporated to dryness under reduced pressure at 55°C on a Buchler Evapomixer, using specially prepared bulb bridges that have ground-glass joints at both ends. No vacum grease of any kind is used on the seals. Samples are taken up and spotted on cellulose thin-layer plates to eliminate interfering substances, which continuously bleed from the AG 1 x 2 resins. Thin-layer chromatography effectively eliminates these interfering substances from the cyclic AMP spot [4]. Reference spots of cyclic AMP (1 μmole) are spotted at each end of the thin-layer plate. Usually 12 samples are spotted on a 20 x 20 cm plate. The plates are developed for a distance of 15 cm in *n*-butyl alcohol-acetic acid-water, 2:1:1. The reference spots of cyclic AMP at each edge are visualized with a short-wavelength UV light, the location of each unknown sample is identified and cut from the plate, and the cellulose is scraped into a polypropylene test tube. Two ml of water is added, the sample is vigorously mixed for 15 sec, and the cellulose is removed by centrifugation. The supernatant is transferred to another 40-ml conical centrifuge tube, then extracted three times with 8 ml of ether. The sample is evaporated to dryness on the evapomixer, the sample is taken up with 30 μl of HCl, pH 2.2, and 20 μl is injected into the high-pressure chromatograph. One sample after another can usually be injected at this concentration of HCl, since compounds more acidic than cyclic AMP would be retained by the column and compounds less acidic have been pre-eluted by use of the column of AG 1 x 2, 50-100 mesh.

B. Use of Cyclic AMP Phosphodiesterase to Partially Destroy Cyclic AMP

Authentic [^{3}H]-cyclic AMP tracer was added to each unknown sample and there was, therefore, the opportunity to verify the authenticity of the cyclic AMP peak in every sample. The specific activity of each sample depends upon the amount of [^{3}H]-tracer cyclic AMP per amount of endogenous cyclic AMP present in the sample tissue extract. An aliquot of the previously purified sample, ready for injection into the chromatograph, was treated with phosphodiesterase to hydrolyze about 50% of the cyclic AMP present in the sample. The treated sample was then injected into the chromatograph. If the cyclic AMP peak was authentic cyclic AMP, then the chromatographic peak height and the radioactivity corresponding to the chromatographic peak should be reduced equally; that is, the specific activity should not change if the peak absorbancy is authentic cyclic AMP [7].

To accomplish this test in practice, a 10 μl fraction remaining from the purified tissue or urine sample is incubated for 10 min at 30°C with 10 μl of a solution containing 5 mM $MgCl_2$, 100 mM tris-HCl, pH 8, and 1 μl of phosphodiesterase [7]. The reaction is stopped in a boiling-water bath for 2 min, and the treated sample is then injected directly into the high-pressure liquid chromatograph. If an equal reduction in peak height and radioactivity of the sample occurs, then one can be reasonably assured that the cyclic AMP peak is authentic cyclic AMP.

V. DETERMINATION OF $[^{14}C]$-CYCLIC AMP SPECIFIC ACTIVITY IN TISSUES PRELABELED WITH $[^{14}C]$-ADENOSINE OR $[^{14}C]$-ADENINE

We were able to demonstrate that this high-pressure chromatographic method can be used to measure [^{14}C]-cyclic AMP specific activity in perfused frog ventricles after the adenine nucleotides were prelabeled with [^{14}C]-adenosine [7]. However, isotopic prelabeling of frog ventricles with [^{3}H]-adenosine has yielded questionable results; i.e., not all of the radioactivity within the cyclic AMP peak was [^{3}H]-cyclic AMP, since treatment with phosphodiesterase caused disproportinate changes in peak height and tritiated material that cochromatographed with the cyclic AMP peak. In addition, measurement of tritium radioactivity in the breakthrough peak and in the chromatographic area following the cyclic AMP peak yielded radioactivity- a finding that caused us to be concerned that some of the radioactivity within the cyclic AMP peak area was not owing to cyclic AMP. On the other hand, measurement of the breakthrough peak and the chromatographic area following the cyclic AMP peak in tissues prelabeled with [^{14}C]-adenosine, yielded no radioactivity - thereby increasing our confidence that the prelabeling method with the use of ^{14}C gave reliable specific activity data for cyclic AMP. The nature of the additional material found with [^{3}H]-adenosine has not been further identified.

VI. CONCLUSION

High-pressure chromatography has proven to be a reliable and versatile method for the measurement of cyclic nucleotides. To date, we have used it to measure tissue and urinary levels of cyclic AMP. In addition, it can be used to measure other cyclic nucleotides, such as cyclic GMP, and cyclic nucleotide analogs, such as 8-iodocyclic AMP and 8-chlorocyclic AMP. The distinctive feature of this methodology is its day-to-day and month-to-month reproducibility and reliability. In addition, no special reagents are necessary to perform this method. Because the final measurement is based upon the inherent ultraviolet absorptivity of cyclic nucleotides, the chromatographic peak heights and areas are directly proportional to the amount of cyclic nucleotide chromatographed. In addition, much reproducibility is gained in tissue and urine analysis because the final result is based upon a specific radioactivity determination.

ACKNOWLEDGMENTS

It is a pleasure to acknowledge the expert assistance of Harrison Frank, Felicidad Avila, and Sharon Laws. Mrs. Georgen Denison provided expert secretarial assistance.

This work was supported by a grant in aid from the American Heart Association, The Los Angeles County Heart Association, the Diabetes Association of Southern California, and the United States Public Health Service, Grant HE 13340.

REFERENCES

[1] C. G. Horvath, B. A. Preiss, and S. R. Lipsky, Anal. Chem., 39, 1422 (1967).

[2] G. Brooker, Anal. Chem., 43, 1095 (1971).

[3] G. Brooker, Anal. Chem., 42, 1108 (1970).

[4] G. Brooker, in Advances in Cyclic Nucleotide Research, (P. Greengard, R. Paoletti, and A. Robison, eds), Vol. 1, Raven Press, New York, in press.

[5] K. Muneyama, R. J. Bauer, D. A. Shuman, R. K. Robins, and L. N. Simon, Biochemistry, 10, 2390 (1971)

[6] O. H. Lowry, N. J. Rosebrough, A. L. Farr, and R. J. Randall, J. Biol. Chem., 193, 265 (1951).

[7] G. Brooker, J. Biol. Chem., 246, 7810 (1971).

Part 2

ENZYMOLOGY OF CYCLIC NUCLEOTIDES

Chapter 4

METHODS FOR THE ISOLATION OF RAT LIVER PLASMA MEMBRANES AND FAT CELL "GHOSTS"; AN ASSAY METHOD FOR ADENYLATE CYCLASE

Martin Rodbell

Section on Membrane Regulation
National Institute of Arthritis
and Metabolic Diseases
Bethesda, Maryland 20014

Adenylate cyclase, an enzyme that catalyzes the conversion of adenosine triphosphate (ATP) to cyclic AMP, is a hormone-sensitive enzyme system that is contained in the plasma membranes of many mammalian cells. The enzyme systems in liver plasma membranes and in fat-cell "ghosts" have proven useful for investigating the mode of action of several peptide hormones and biogenic amines [1,2]. In this chapter, methods will be described for purifying liver plasma membranes and fat-cell ghosts. The procedures used in this laboratory for assaying adenylate cyclase also will be detailed.

I. PURIFICATION OF LIVER PLASMA MEMBRANES (according to the procedure of Neville [3])

A. Solutions

1. "Medium": 1 mM $NaHCO_3$: At least two liters are required; prepare in advance and chill in cold room overnight.

2. 69% Sucrose: Weigh out 69 g of sucrose, add 31 ml of water, and heat on hot plate until sucrose dissolves. After cooling, adjust to 69 $\pm$ 0.5%, as measured by an Abbe refractometer, by the addition of solid sucrose or water as needed to raise or lower the concentration. Place solution in ice bucket.

3. 42.3% Sucrose: Weigh out 42 g of sucrose, add 57 ml water, and heat on hot plate until sucrose dissolves. After cooling, adjust the solution carefully to 42.3% $\pm$ 0.1% by the addition of 69% sucrose or water to raise or lower the concentration. Place in ice bucket.

NOTE: All tissue, media, sucrose solutions, etc., must be kept cold in ice bath at all steps of the procedure.

B. Equipment

1. Dounce loose-fitting homogenizer; obtain

large homogenizer from Blaessig Glass Specialties, Rochester, N. Y. The pestle and homogenizer should have a tolerance such that the plunger falls to the bottom of the water-filled homogenizer in 5 sec.

2. Refrigerated centrifuge (International type) kept at 0°-5°C.

3. Beckman preparative centrifuge (or suitable high-speed preparative centrifuge) and SW 25.2 rotor, both previously cooled to 0°-4°C.

4. Abbe refractometer.

C. Method

1. Decapitate enough rats (usually 12 to 16) to give 80 g of liver. Young female rats (120-150 g) are generally preferred. Quickly remove the livers and place them in an iced beaker.

2. Weigh out 10 g of liver in a small plastic cup and mince with scissors. Remove any connective tissue while mincing.

3. Place the minced liver in the Dounce homogenizer. Add 25 ml of medium. Homogenize in the ice bucket with eight vigorous strokes of the pestle.

4. Add the homogenate to 450 ml of cold medium in a 1-liter beaker.

5. Repeat steps 2 and 3, and add homogenate to the same medium as in step 4. Stir for 3 min (by a

timer). Filter the mixture first through two layers, then four layers of cheesecloth.

6. Repeat steps 1 to 5.

7. Decant the two batches of filtered homogenate into four 250-ml centrifuge bottles and centrifuge for 30 min at 2650 rpm (1500 x g) in the refrigerated centrifuge at 4°C.

8. Decant and discard the supernatant fluid, removing as much fluid as possible. Pour the pellets into clean Dounce homogenizer kept in an ice bucket.

9. Repeat steps 1 to 8, adding the pellets to the same homogenizer. Resuspend the pellets with three gentle strokes of the loose pestle.

10. Decant the resuspended pellets into a 250-ml graduated cylinder with a ground-glass stopper that contains 62 ml of 69% sucrose (adjusted during the time of centrifugation in step 7). Add water to make 110 ml. Mix thoroughly and keep cold. Adjust concentration to 44.0 ± 0.1% using 69% sucrose or water. If 69% sucrose has been carefully adjusted, usually little further adjustment is required to achieve a concentration of 44% sucrose.

11. Distribute the sucrose suspension equally among three 1-1/4" x 3-1/2" cellulose nitrate centrifuge tubes. Carefully overlay each tube with 20 ml

of 42.3% sucrose.

12. Centrifuge for 150 min at 25,000 rpm in an SW 25.2 rotor at 4°C. Handle the tubes carefully to preserve the density interface; grease the caps well.

13. Remove the floated material with a spoon-shaped spatula and place it in a plastic centrifuge tube. Add about 10 ml of medium and resuspend by drawing the entire contents of the tube through a #20-gauge needle into a large syringe and expelling the mixture back into the tube. Fill the tube with medium and centrifuge for 15 min at 15,000 rpm in a refrigerated centrifuge (Sorvall type).

14. Carefully decant and discard the supernatant fluid. The pellet consists of "partially purified" plasma membranes.

NOTE: "Partially purified" plasma membranes ordinarily are used in this laboratory for investigation of the effects of glucagon on adenylate cyclase A procedure for further purifying the plasma membranes is described in the remaining steps, but it results in lower yields of enzyme, without any change in specific activity of the enzyme and its response to glucagon.

15. Add 4 ml of medium to the pellet of step 14 and resuspend the pellet as in step 13.

16. Into the bottom of each of two siliconized S-25 centrifuge tubes, place a "cushion" of 4.1 ml of 50 $\pm$ 1% sucrose. Over each cushion form a linear sucrose gradient from 24-1%.

17. Overlay gradients with 2 ml of resuspended homogenate, and centrifuge at 2000 rpm (550 x g, max) in a Spinco for 1 hr with the brake off.

18. By use of a syringe and a long, blunt #20 needle, remove the material at the cushion interface. This material represents "purified" plasma membranes.

NOTE: "Partially" or "fully" purified plasma membranes from steps 14 or 18 are frozen immediately and stored under liquid nitrogen. Under frozen conditions, adenylate cyclase activity and response to glucagon are retained after months of storage. The pellets can be suspended also in a sodium bicarbonate medium, distributed in aliquots of 0.2-0.5 ml, and frozen as above; this procedure has the advantage of not requiring that an entire batch of membranes be thawed for each assay.

II. ISOLATION OF FAT CELLS FROM RAT ADIPOSE TISSUE

A. Reagents and Materials

1. Collagenase: Crude enzyme (120-150 units/mg) obtained from Worthington Biochemical Corp. (Freehold,

New Jersey).

2. Albumin: Bovine serum albumin (fraction V), obtained from commercial sources (Armour Pharmaceuticals, Miles Laboratories).

3. Silk Screen: Mesh size, 7 x 10x, obtained from artist's supply houses.

4. Plastic Vials: Volume, 25 ml, obtained from various supply houses.

5. Plastic Centrifuge Tubes: Volume, 12-15 ml, obtained from various supply houses.

6. Plastic Syringe Filtering Device: Cut off the bottom portion of a 25-ml plastic syringe. Attach silk screen to the end of the syringe with a band of heavy rubber tubing.

7. Whatman #1 Filter Paper (5.5 cm Disks).

NOTE: Plastic containers are used for the preparation of fat cells, since fat cells tend to break when in contact with glass.

B. Equipment

1. Table top centrifuge.

2. Mickle tissue slicer (H. Mickle, Surrey, England);optional.

3. Shaking water bath, adjusted for 15 cycles per minute at 37°-38°C.

C. Solutions

Incubation or Washing Medium

Reagent	Volume, ml	Final molarity, mM
NaCl (2.5 M)	10.0	125
KCl (1.0 M)	1.0	5
$CaCl_2$ (0.05 M)	4.0	1
$MgCl_2$ (0.05 M)	10.0	2.5
KH_2PO_4 (1.0 M)	0.2	1
Tris-HCl, pH 7.4 (1.0 M)	5.0	25
4 g albumin	---	2%

Add water to make 200 ml. Make sure that the pH is 7.4-7.6. If it is not, adjust with either 2 M HCl or 1 M tris base.

D. Procedure

The following procedure is for 8 rats (male or female) maintained on a normal diet and weighing from 150 to 250 g. Fasted animals and animals weighing less that 150 g yield fewer fat cells and fat-cell ghosts.

1. Kill rats by decapitation or cervical dislocation. Quickly remove epididymal (male) or parametrial (female) fat pads and place them in the incubation medium at room temperature.

2. Place pads from one animal on filter paper wetted with the incubation medium. Slice the tissue with a Mickle slicer. Alternatively, the tissue, wetted with the incubation medium, may be minced with scissors.

3. Transfer the minced tissue to a plastic vial containing 3 ml of incubation medium plus collagenase.

NOTE: Since collagenase varies in potency, trial experiments should be carried out to determine the concentration necessary to cause disaggregation of tissue in 1 hr; this concentration varies between 0.5 and 1.0 mg of collagenase per ml.

4. Incubate tissue for 1 hr. (This time period is convenient for preparing incubation mixtures for adenylate cyclase assays.)

5. With the aid of a syringe plunger, express the contents of each vial through a silk screen. This procedure causes dispersion of clumped cells and removes undigested tissue.

NOTE: To prevent clogging of the silk screen, only one screen should be used for two vials of digested adipose tissue.

Collect the cells in a plastic container; scrape off cells clinging to the outer surface of the screen and add to the plastic container.

6. Distribute the combined cell suspensions into four plastic centrifuge tubes. Centrifuge for just the time necessary to reach 1000 rpm (about 15 sec). This causes the fat cells to float rapidly to the surface.

7. Aspirate the infranatant fluid (use a Pasteur pipet attached with rubber tubing to a water aspirator) and suspend the fat cells, by gentle shaking, in 5 ml of fresh incubation medium.

8. Repeat steps 6 and 7 twice more for a total of three washings.

9. After the third wash, combine the suspended fat cells and pass them (without use of a plunger) through the silk screen. This step removes essentially all remaining stromal-vascular cells.

10. Distribute the cells equally among four plastic centrifuge tubes and centrifuge them as in step 6. Remove the infranatant fluid. Fat cells are now ready for lysis and preparation of fat-cell ghosts.

III. PREPARATION OF FAT-CELL GHOSTS

Ghosts of fat cells are sacs of plasma membrane within which are contained mitochondria, endoplasmic reticulum, and, occasionally, a nucleus [2]. The procedure described in subsection B is based on swelling of the fat cells with a hypotonic medium containing salts (particularly Ca^{2+} and Mg^{2+}) which serve to stabilize the membranes, and ATP, which seems to stabilize adenylate cyclase and its response to hormones. Sulfhydryl agents (mercaptoethanol) are added only because adenylate cyclase is a sulfhydryl-containing enzyme. Hypotonic swelling alone does not suffice to cause rupture or lysis of the cells. Agitation of the swollen cells is necessary and is carried out in a manner that produces relatively large sacs that are readily sedimented at low gravitational forces. Because adenylate cyclase is an unstable enzyme (even when kept at low temperatures), it is essential that the ghosts be prepared as rapidly as possible and assayed within an hour after their isolation.

A. Solutions

1. Lysing Medium

Reagent	Volume (ml/50 ml)	Final molarity (mM)
$MgCl_2$ (0.05M)	2.5	2.5
ATP (0.05 M)	2.5	2.5
$CaCl_2$ (0.05 M)	0.1	0.1
$KHCO_3$ (0.1 M)	0.5	1.0
tris-HCl, pH 7.6 (1 M)	0.1	2.0
mercaptoethanol (0.1 M)	0.5	1.0

Lysing medium should be prepared freshly and kept in an ice bath. The inclusion of mercaptoethanol in the lysing medium is optional.

2. Suspending Medium

The suspending medium is a solution of 1.0 mM $KHCO_3$ containing 1.0 mM mercaptoethanol. This solution should be prepared fresh for each experiment and kept in an ice bath.

B. Procedure

1. Add 5 ml of lysing medium to each centrifuge tube containing washed fat cells (step 10, isolation of fat cells). Gently mix contents until cells are well dispersed.

2. Centrifuge for 15 sec in a table-top centrifuge. Aspirate the infranatant fluid. This step is designed to wash fat cells free of the incubation medium and to cause swelling of the fat cells in the hypotonic medium.

3. Repeat step 1, but now invert the tubes in a cyclical fashion 20 times, using a vigorous shaking motion at each downward stroke. This step causes breakage of the fat cells.

4. Centrifuge in a table-top centrifuge for about 30 sec.

5. With the aid of a Pasteur pipet, remove the cloudy infranatant and pellet and transfer them to a glass centrifuge tube (13 ml, conical) chilled in an ice bath.

6. Repeat steps 3 and 4; combine infranatant fluids and pellets.

7. Centrifuge the combined lysates for 15 min at 900 x g (2000 rpm in International Centrifuge, rotor 269), at 4°C.

8. Aspirate the supernatant fluid. Suspend the pellet, with the aid of a wide-mouthed pipet (plastic preferred), in 5 ml of suspending medium.

NOTE: Do not disperse the pellet with a narrow-bore pipet, since the resultant shearing forces may cause formation of smaller vesicles of plasma membranes with consequent lower yields of ghosts.

9. Centrifuge as in step 7. Suspend the pellet in a suspension medium equal to 5 times the volume of the pellet. The pellet volume obtained from lysis of fat cells from 8 rats is usually about 0.3 ml. Dilutic to 1.5 ml results in a concentration of membrane protein of about 3.5 mg/ml.

NOTE: Ghosts should be kept in an ice bath, and should be assayed as quickly as possible for adenylate cyclase activity; activity and response to hormones decays by about 50% during storage for 4 hr at 0°C.

IV. KRISHNA METHOD FOR ASSAYING ADENYLATE CYCLASE ACTIVITY

Essential to any method for the assay of adenylate cyclase activity is the ability to separate labeled cyclic AMP completely from all other labeled nucleotides present or formed during the incubation of the enzyme system with appropriately labeled ATP. The method developed by Krishna [4] satisfies this criterion and takes advantage of the following behavior of cyclic AMP: the nucleotide, unlike ATP, ADP, and P_i, is retained on Dowex-50 (H^+) columns, and cyclic AMP is not adsorbed, unlike other nucleotides and P_i, to nascently formed $BaSO_4$. Because cyclic AMP can be formed nonenzymatically from ATP, particularly under slightly alkaline conditions, it is necessary first to separate the bulk of the ATP from cyclic AMP on Dowex-50 columns. The original procedure [4] used a Dowex-50 chromatographic separation, followed by two consecutive precipitations with small, and equal volumes of $ZnSO_4$ and $Ba(OH)_2$, followed each time by centrifugation. This procedure is still probably the most widely used version of this technique. The procedures used in our laboratory permit the assay of 200 samples by one operator within four hours, with less than 1%

variation between duplicate samples.

A. Reagents and Solutions

1. Radioactive

a. [α-^{32}P]-ATP, specific activity, 2-4 Ci/mmole.

b. [^{3}H]-Cyclic AMP (2-5 Ci/mmole), diluted with water to contain between 300,000 and 500,000 DPM/ml.

2. Na_2ATP (25 mM).

NOTE: The solution should be adjusted to about pH 7.0 before use; store frozen.

3. Creatine-phosphate (200 mM) containing 10 mg/m of creatine kinase (Sigma) as an ATP regenerating system. This solution is used for maintaining a constant concentration of ATP during incubation with liver and fat-cell membranes, which contain potent ATPases. Disodium creatine-phosphate from Sigma is purified and converted to its tris salt by dissolving 750 mg in 4 ml of water and placing the solution over a column (1 x 20 cm) of Dowex-50 (H^+) (see item 8) and discarding the eluate. Add 8 ml of water and collect the eluate. Immediately adjust the pH to 7.6 with 2 M tris. Dilute to 10 ml and add 100 mg of creatine kinase. Store the solution frozen.

4. Tris-HCl, pH 7.6 (0.5 M).

5. $MgCl_2$, 125 mM.

6. Cyclic AMP, 50 mM.

7. "Stopping solution". Dissolve 850 mg of cyclic AMP, 5.0 g of $Na_2ATP \cdot 4\ H_2O$, and 2.0 g of sodium dodecyl sulfate in 180 ml of water. Adjust the pH to 7.6 with 2 M tris; dilute to 200 ml. Keep the solution frozen until ready for use. The detergent in this solution immediately and completely stops adenylate cyclase activity. The ATP and cyclic AMP are added as "carriers" respectively to diminish any breakdown of labeled ATP to cyclic AMP and to reduce loss of labeled cyclic AMP during isolation.

8. 50% (w/v) Dowex-50 (H^+). Weigh out 500 g of Bio-Rad AG 50W-X8 and add to a 1-liter graduated cylinder. Add water to the 1-liter mark. Mix thoroughly and allow the resin to settle; decant the supernatant fluid. Repeat this operation until the supernatant is colorless. Add water to make 1 liter and transfer the mixture to a 1-liter flask containing a large magnetic stirring bar. Store the flask at room temperature.

9. $Ba(OH)_2$ (0.3 N). Store this solution in a tightly stoppered bottle.

10. $ZnSO_4$ (0.3 M).

NOTE: Solutions 9 and 10 are used to form nascent $BaSO_4$. For maximal adsorption of nucleotides it is essential that the pH of the final mixture be between 7.5 and 8.0. Titrate the solutions with a pH meter and adjust concentrations such that the addition of equal volumes gives the desired pH.

11. Aquasol (New England Nuclear). This solution is used for scintillation counting. Aquasol has the advantage over most other scintillation solutions of forming stable gels, with minimal quenching, in the presence of 3 ml of water.

B. Standard Medium for Adenylate Cyclase Assays

Basic Assay Cocktail.

Solutions are prepared so that the addition of 20 μl of cocktail to the assay medium (50 μl) will give final concentrations of the reagents as indicated.

Reagent	Volume (ml/1 ml)	Final concentration
$MgCl_2$	0.1	5 mM
ATP	0.1	1 mM
Cyclic AMP	0.1	2 mM
ATP-Regenerating Mixture	0.25	20 mM creatine-P + 1 mg/ml creatine kinase
Tris-HCl, pH 7.6	0.1	20 mM
[α-^{32}P]-ATP	Add 100-150 cpm/pmole of ATP	
H_2O	To make 1 ml final volume	

C. Incubation Procedure

1. Add 20 μl of assay cocktail to 10 x 75 mm glass test tubes kept in an ice bath. Add hormones diluted in 0.5% albumin to the desired concentrations. Controls must always contain an equivalent concentration of albumin. Add 10- or 20-μl suspensions of liver plasma membranes or fat-cell ghosts, to make a final volume of 50 μl.

2. Initiate reactions by adding enzyme, mixing the total on a Vortex mixer, and adding each tube, at 15-sec intervals to a water bath at 30°C.

3. Reactions are stopped at appropriate time intervals by the addition of 0.1 ml of stopping solution.

<u>NOTE</u>: Zero-time controls are prepared by the adddition of stopping solution before the addition of ghosts or liver membranes. Zero-time controls should contain the same reaction mixtures used in the experiment. One zero-time control for each variant is sufficient.

4. Add 50 μl of [^{3}H]-cyclic AMP solution. This solution should be added as accurately as possible, since the over-all recovery of [^{32}P]-cyclic AMP formed during incubation is based on the recovery of [^{3}H]-

cyclic AMP.

5. Upon completion of a rack of tubes, place the rack in a tray of boiling water for five minutes.

6. Add 0.9 ml of water to each tube.

D. Isolation of Cyclic AMP

1. Dowex-50 Chromatography

a. Add 2.0 ml of Dowex-50 suspension (rapidly stirred) to either glass (0.4 x 8 cm) or plastic (Chromaflex, Kontes) columns containing a small wad of glass wool. The columns are suspended on a perspex rack containing suitably spaced holes to hold 50 columns. Rows of thermometer clamps are also satisfactory for holding the columns.

b. Decant the reaction tubes (step 6) into the tops of the columns; be sure that the contents drain into the resin.

c. Discard the eluate. With a syringe, add 1.25 ml of distilled water. Allow to drain and add another 1. 25 ml of water. Discard the eluates.

NOTE: Collect the eluates into a porcelain tray. Since this tray contains a large amount of radioactiv solutions should be disposed of by decantation into a large plastic bottle which should be shielded.

d. Replace the tray with a test-tube rack

designed to hold 50 13 x 100-mm glass test tubes positioned directly under the tips of the columns.

e. Add 3 ml of distilled water to each column. Collect the eluate, which contains labeled cyclic AMP.

2. Treatment with Nascent Barium Sulfate

Add, successively, to each tube 0.1 ml each of $ZnSO_4$ and $Ba(OH)_2$ solutions. Mix immediately and thoroughly on a Vortex mixer.

3. Filtration of Barium Sulfate Suspensions

Insert a small wad of glass wool in the bottom of glass (0.4 x 8 cm) or plastic (Chromaflex, Kontes) columns. Using syringes, add simultaneously 0.2 ml of $Ba(OH)_2$ and $ZnSO_4$ solutions to make a plug of nascently formed $BaSO_4$. This plug serves to filter the $BaSO_4$-containing, adsorbed labeled nucleotides (step 2 of this subsection) and allows the solution of labeled cyclic AMP to be separated rapidly and cleanly from $BaSO_4$. The columns are placed in the same type of perspex rack described in step 1a of this subsection, containing 50 holes, and are positioned such that each column is directly over a set of 50 counting vials in a suitably constructed rack.

Invert the test tubes from step 2 into the columns and allow the contents to drain into the

the counting vials which had been previously filled with 12 ml of counting solution (Aquasol). After the fluid has completely drained into the vials, shake the vials vigorously and place them into a scintillation counter.

E. Counting Procedure

Samples are counted in a Packard Tri-Carb (model 3320), or any other type of 2-channel scintillation counter. In the Packard counter, the channels are adjusted as follows:

Tritium	^{32}P
Window A-B	Window E-infinity
50% gain	2% gain
40-500	230-1000

With these adjustments, no significant tritium counts appear in the ^{32}P channel, and less than 2% of the ^{32}P counts, appear in the tritium channel. The 2% spillover of ^{32}P is inisignificant when we consider that the [^{3}H]-cyclic AMP activity is always at least 10 times that found in [^{32}P]-cyclic AMP.

F. Calculations

$$\frac{\text{Cyclic AMP } ^3\text{H (standard)}}{\text{Cyclic AMP } ^3\text{H (in sample)}} \times \frac{^{32}\text{P in sample} - {}^{32}\text{P background}}{\text{specific activity of } [^{32}\text{P}]\text{-ATP}} =$$

pmoles of cyclic AMP

(pmoles of cyclic AMP formed)-(zero-time cyclic AMP) =

net production of cyclic AMP.

NOTE: Zero-time cyclic AMP represents ^{32}P-labeled material present in the [^{32}P]-ATP that behaves as cyclic AMP through the isolation procedure. In some commercial preparations, this background may be unsatisfactorily high; i.e., greater than 0.005%. If this proves to be the case, the contamination of contaminants can be reduced by passage of the solution of high specific activity [^{32}P]-ATP through a small column of Dowex-50 (H^+), as in step 1 a of section IV, D and collection of the eluate.

Specific activity of ATP is determined by measurement of the absorbance at 260 nm of an appropriately diluted (usually 1 to 500 if 1 mM or higher) sample of the incubation cocktail. Use an extinction coefficient of 15,000. Diluted samples are counted under the same conditions (Aquasol, 3 ml of water) as are the samples described herein.

The concentration of membrane protein ("ghosts" or liver plasma membranes) added to the incubation tubes is determined by the methods of Lowry et al. [5] and Layne [6]. Activities of adenylate cyclase should be expressed in pmoles (or nmoles) of cyclic AMP formed per min per mg of membrane protein.

REFERENCES

[1] M. Rodbell, A. B. Jones, G. E. Chiappe de Cingolani, and L. Birnbaumer, Recent Prog. Hormone Res. 24, 215, (1968).

[2] S. L. Pohl, L. Birnbaumer, and M. Rodbell, J. Biol. Chem., 246, 1849, (1971).

[3] D. M. Neville, Biochim. Biophys. Acta, 154, 540, (1968).

[4] G. Krishna, B. Weiss, and B. B. Brodie, J. Pharm Exptl. Therap., 163, 379, (1968).

[5] O. H. Lowry, N. J. Rosebrough, A. L. Farr, and R. J. Randall, J. Biol. Chem., 193, 265, (1951).

[6] E. Layne, Methods Enzymol., 3, 447, (1957).

Chapter 5

GUANYL CYCLASE
PARTIAL PURIFICATION AND ASSAY

Arnold A. White, Sharon J. Northup and Terry V. Zenser

Space Sciences Research Center and
Department of Biochemistry
University of Missouri
Columbia, Missouri 65201

I. INTRODUCTION

Cyclic 3',5'-guanosine monophosphate was first identified in nature by Ashman et al. [1], who administered ^{32}P inorganic phosphate intraperitoneally into rats and 24 hr later identified by paper chromatography the labeled organophosphate compounds present in the urine. The only nucleotides found were cyclic GMP and cyclic 3',5'-adenosine monophosphate. Cyclic AMP had been identified earlier by Sutherland and Rall [2]. Price et al. [3] further characterized cyclic GMP after large scale isolation from rat urine and suggested that it was synthesized from guanosine triphosphate (GTP) by a reaction analogous to that catalyzed by adenyl cyclase (see Chapter 4).

The enzyme involved in the formation of cyclic GMP, guanyl cyclase, was detected nearly simultaneously by three groups. Bohme et al. [4] showed that various rat tissues could convert [^{14}C]-GTP into [^{14}C]-cyclic GMP. Hardman and Sutherland [5] measured the formation of cyclic GMP by hydrolyzing the purified nucleotide with cyclic nucleotide phosphodiesterase. The product, 5'-GMP, was determined by the use of an enzymatic recycling technique. Hardman and Sutherland also used an assay based on the conversion of [^{3}H]-GTP to [^{3}H]-cyclic GMP. White and Aurbach [6] based their assay for guanyl cyclase on the use of [α -^{32}P]-GTP. Because of the specific labeling of the substrate, its conversion into [^{32}P]-cyclic GMP was one proof that the guanyl cyclase reaction occurred as originally suggested by Price et al. [3]. The work discussed here uses this substrate exclusively.

II. DEVELOPMENT OF THE ASSAY

In our original study on the guanyl cyclase activity of rat tissues, we concluded that the relative activity of the enzyme in these tissues differed [6]. Thus, we found lung to have the highest activity, with brain and spleen having lower activities, followed by kidney, liver, and heart, which were about equally active. However, when we began to apply this assay to measure the activity of the same tissue from different animals, we found considerable variation. While this could have been due to differences in age or physiology, we began to suspect that our assay was not reliable. Our first indications of the cause of the variation were the results of the experiment shown in Fig. 1. The specific activity of each tissue is plotted as a function of the amount of the tissue supernatant that was assayed. Under these assay conditions, the measurable activity of certain tissues was very markedly affected by the concentration

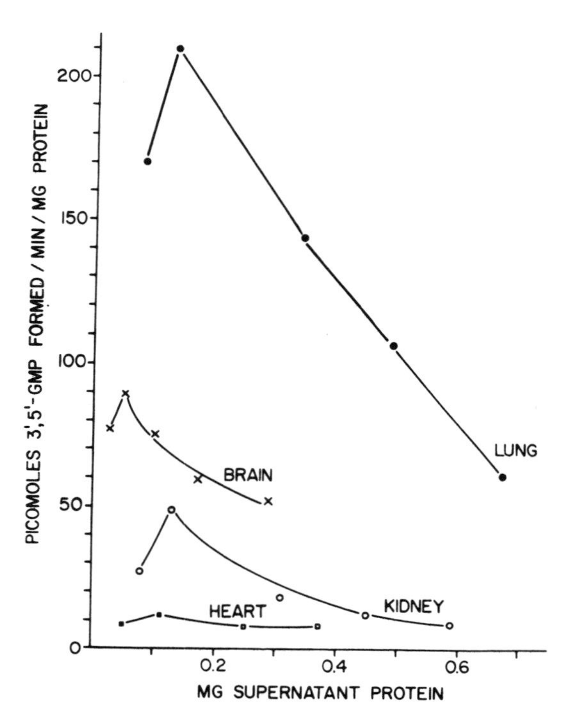

Fig. 1. Guanyl cyclase specific activity of 37,000 x g supernatants prepared from rat tissues, as affected by amount of supernatant protein assayed. Reaction mixture contained 0.4 mM GTP, 10 mM $MnCl_2$, 20 mM caffeine, 50 mM tris-HCl, pH 7.6, 15 μg bovine plasma albumin, and 2.67 mM cyclic GMP. (Brain was assayed in the presence of 5.34 mM cyclic GMP). Incubations were for 10 min at 37°C. The radioactive cyclic GMP formed was determined by Method A (Section II.B.4.a).

of the tissue supernatant fraction used as the enzyme in the reaction mixture.

The assays diagramed in Fig. 1 were performed at 37°C, with a 10-min incubation time. We later found that we could improve the results considerably (that is, flatten the curves) by shortening the incubation time. This suggested that the difficulties encountered with a 10-min incubation time were due to a limited supply of substrate or a rapid destruction of the product, or both. Since we had a high concentration of caffeine in the reaction mixture as a phosphodiesterase inhibitor, and in addition a large amount of cold cyclic GMP to act as a trap, we were doubtful that we could further decrease interference by phosphodiesterase. We decided to give our attention to the problem of enzymes which could destroy GTP(GTPase).

In Fig. 2 are the results of an experiment designed to determine if the problem was indeed competing GTPase. GTPase in a 37,000 x g supernatant from a rat liver homogenate is most probably associated with ribosomes [7]. We had earlier found guanyl cyclase activity in a 104,000 x g supernatant from liver [6]; such a supernatant would be relatively free from ribosomal GTPase. To insure removal of GTPase we centrifuged a 37,000 x g supernatant for 2 hr at

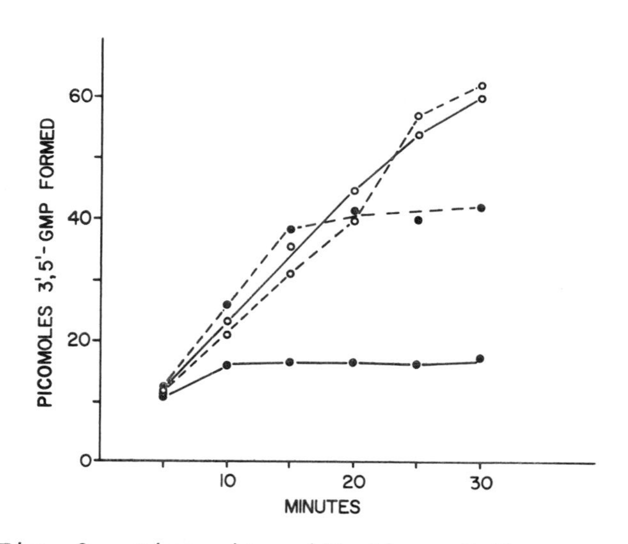

Fig. 2. Linearity with time of the guanyl cyclase activity of rat liver supernatants. The liver from a 180-g male rat was perfused _in situ_ with buffer and homgenized in 3 volumes (w/v) of 0.05 M tris-HCl, pH 7.6, containing 50 mM mercaptoethanol. The homogenate was centrifuged at 37,00 x g for 30 min. Part of the supernatant was retained for assay and the rest was centrifuged for 2 hr at 198,00 x g (max). Guanyl cyclase activity was determined on equal volumes (10 µl) of each supernatant, there being used 450 µg protein of the 37,000 x g supernatant (closed circles) and 265 µg protein of the 198,000 x g supernatant (open circles). Assay was at 30°C, in 0.05 M tris-HCl, pH 7.6, 20 mM caffeine, 0.4 mM GTP,

198,000 x g. Figure 2 shows that such a high speed supernatant did result in a guanyl cyclase reaction linear for a much longer time than did the same volume of the 37,000 x g supernatant. This diagram also shows that the addition of a GTP regenerating system to the 37,000 x g supernatant resulted in a reaction linear for 15 min, while addition to the high speed supernatant had no effect. This latter result supports the view that GTPase interferes with the guanyl cyclase assay, and its removal improves the assay.

Alternatively, shorter incubation times, lower incubation temperatures, increased GTP concentrations, and an effective GTP-regenerating system would also improve the assay by decreasing interference by GTPase. Hardman and Sutherland [5] found the K_m for GTP of rat lung guanyl cyclase to be between 0.02 and 0.1 mM, in the presence of 3 mM $MnCl_2$. Northup [8]

2.67 mM cyclic GMP, 1 mM $MnCl_2$, and 15 μg bovine plasma albumin. Assay of both supernatants was also performed in the presence of a GTP-regenerating system (interrupted line) which consisted of 50 mM KCl, 1 mM phosphoenolpyruvate, and 5 μg pyruvate kinase. The reaction mixtures were analyzed by Method B (section II.B.4.b).

determined that the Km for GTP of several rat tissues (in the presence of 6 mM $MnCl_2$) varied from 0.08 to 0.37 mM. Since GTP in excess of 1.2 mM was inhibitory [8], we decided to work with 1.2 mM GTP. This was about 3 to 4 times the highest Km we expected to encounter.

The experiment depicted in Fig. 1 was repeated with the new assay conditions (6 mM $MnCl_2$, 1.2 mM GTP, and a 5-min reaction time at 30°C); the results are shown in Fig. 3. There was an obvious improvement in the results of the assay, as compared to Fig. 1, for all tissues except brain and kidney. Since brain contains very high phosphodiesterase activity, the poor results with brain depicted in Fig. 3 were ascribed to too low a "catcher" pool of cyclic GMP. However, the results with respect to kidney could not be explained on this basis.

Assays showed that the guanyl cyclase preparations contained high amounts of GTPase, inhibited only slightly by NaF. The effect of this GTPase on the guanyl cyclase reaction could be minimized by including a GTP-regenerating system. Since PEP inhibits guanyl cyclase[5], but we have found that creatine phosphate does not, (Table 1), we decided to incorporate a creati phosphokinase system into the guanyl cyclase reaction mixture. Creatine phosphate was used at 5 mM, which was not inhibitory.

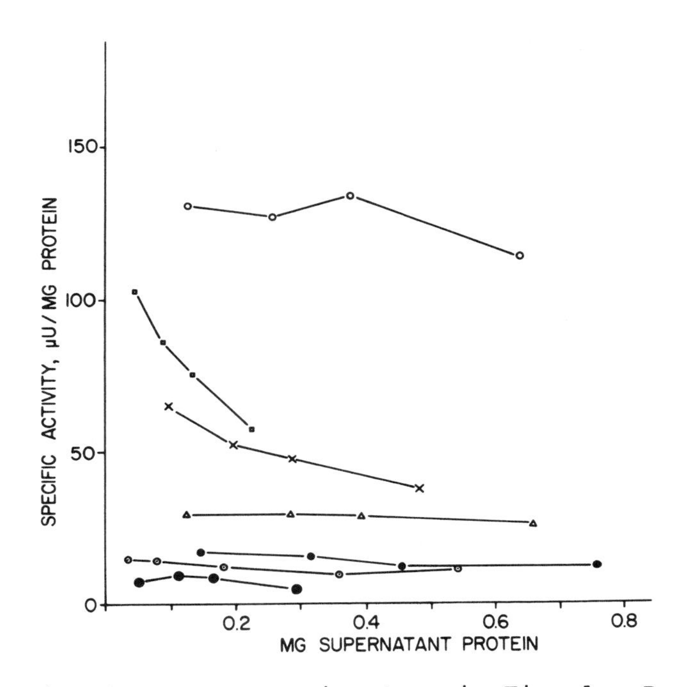

Fig. 3. Same experiment as in Fig. 1. Reaction mixture changed to contain 1.2 mM GTP, 6.0 mM $MnCl_2$, 20 mM caffeine, 50 mM tris-HCl, pH 7.6, 0.1 mg bovine plasma albumin, and 2.67 mM cyclic GMP. Incubations for 5 min at 30°C. The rat tissues assayed were lung (-O-), brain (-□-), kidney (-X-), spleen (-Δ-), liver (-•-), heart (-⊙-), and gastrocnemius muscle (-●-).

TABLE 1

Effect of Creatine Phosphate and Phosphoenolpyruvate on Guanyl Cyclase Activity

Creatine Phosphate (mM)[a]	Inhibition (%)	Phosphoenol-pyruvate (mM)[b]	Inhibition (%)
0	0	0	0
5	0	1	15
10	2	2	22
12.5	5	4	29
15	6	8	37
25	18	10	43

[a]An ammonium sulfate fraction prepared from rat lung was used for these experiments. It was dialyzed overnight against 0.05 M tris-HCl containing 10 mM mercaptoethanol, and centrifuged briefly to remove a precipitate; it was stored at -20°C. Ten μl contained 39 μg of protein. Guanyl cyclase activity was determined in a standard reaction mixture, without a GTP regenerating system.

[b]An ammonium sulfate fraction prepared from rat kidney was used for these experiments. It was dialyzed for 2 hr against 0.05 M tris-HCl containing 50 mM mercaptoethanol. Ten μl of this preparation contained 118 μg of protein. Guanyl cyclase was determined as described above for the lung preparation.

Figure 4 shows the results of the first test of the new regenerating system. While there was not as profound an effect with the kidney supernatant as we would expect, there was an increase in measurable activity with the lung preparation even at the 5-min reaction time. The linearity of the reactions with time was clear for both tissues. Large amounts of supernatant protein were used, yet the reactions were linear for 10 min with kidney and for 15 min with lung.

As a final test of the GTP regenerating systems, we again studied specific activity as affected by supernatant concentration. Figure 5 shows the results obtained with creatine phosphokinase and creatine phosphate. The regenerating system improves results with lung, brain, and kidney. It appears that the supernatant protein concentration that appears most reliable is in the range of 0.2 to 0.3 mg per reaction mixture. The following methods incorporate the improvements in the guanyl cyclase assay that have resulted from these studies.

III. METHODS

A. Preparation of Guanyl Cyclase

1. Extraction of Tissues

In some of the work described, tissue homogenates were prepared in 0.05 M tris-HCl, pH 7.6. Since guanyl

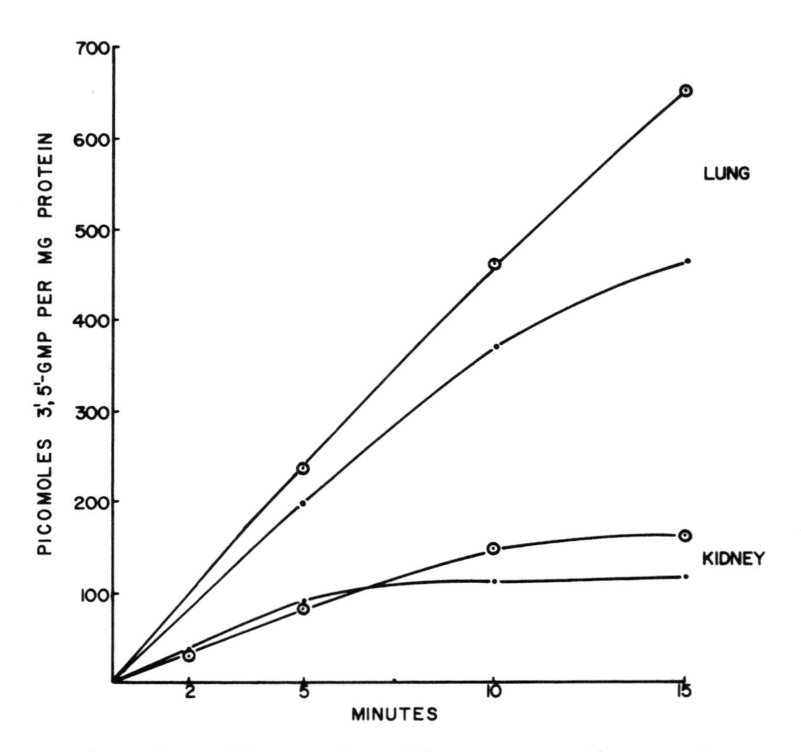

Fig. 4. Effect of a GTP-regenerating system on guanyl cyclase assay. Rat lung and kidney were homogenized with three parts (w/v) of 0.05 M tris-HCl, pH 7.6, containing 0.5 mM EDTA and 1 mM dithiothreitol. A 37,000 x g supernatant was prepared, which contained 0.592 mg of lung protein and 0.498 mg of kidney protein in the 25 µl of each that was assayed. A standard incubation mixture was used, both with (-⊙-) and without (-•-) a GTP-regenerating system (15 µg creatine phosphokinase plus 5 mM creatine phosphate).

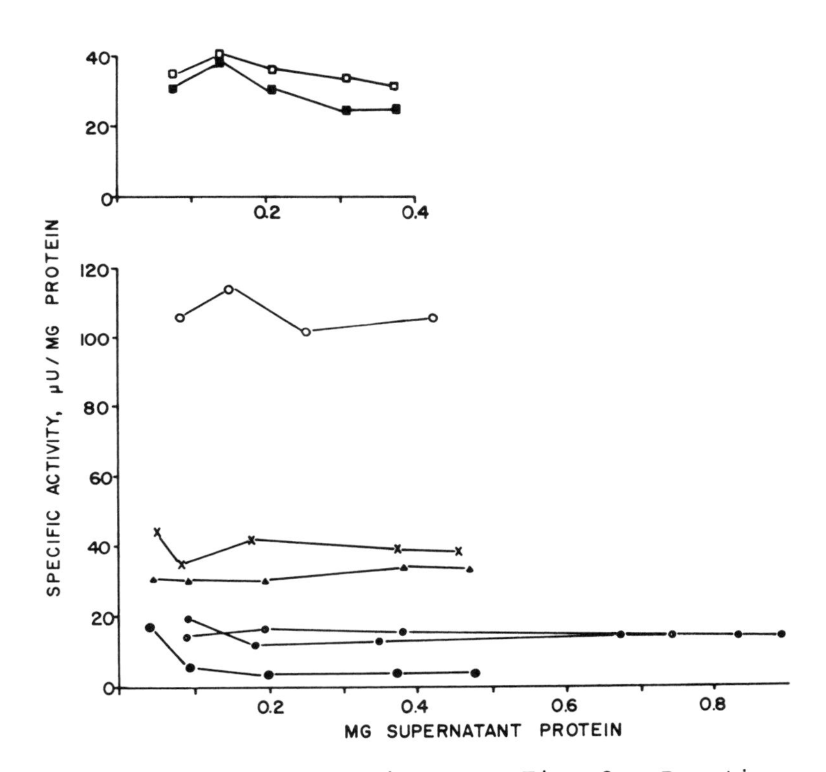

Fig. 5. Same experiment as Fig. 3. Reaction mixture modified by the addition of 10 mM mercaptoethanol and of a GTP-regenerating system consisting of 15 μg creatine phosphokinase and 5 mM creatine phosphate. Brain was assayed with both 2.67 mM (-■-) and 5.34 mM cyclic GMP (-□-).

cyclase is a sulfhydryl-containing system, either 10 mM mercaptoethanol or 1 mM dithiothreitol is now added routinely. (In addition, we usually add 0.5 mM EDTA in order to remove traces of heavy metals.) Glass-distilled, deionized water must be used for all solutio

Tissues are cut with scissors and homogenized with three vol of buffer (w/v) in a Potter-Elvehjem type homogenizer. A Kontes ground-glass homogenizer (Duall) is used for heart, lung, and skeletal muscle. A smooth glass tube with a Teflon pestle is used for other tissues. The homogenates are centrifuged for 30 min at 37,000 x g (max) in a Sorvall refrigerated centrifug Model RC-2, and the supernatant solution assayed for the enzymatic activity.

2. Preparation of Ammonium Sulfate Fraction

We have previously shown that guanyl cyclase precipitates from tissue extracts between 20 and 40% of saturation with ammonium sulfate [6]. This fraction is comparatively stable when frozen and has nearly all of the GTPase activity removed. It is, therefore, useful in order to study the effects of certain compoun on guanyl cyclase activity.

The tissue to be processed is homogenized in three vol of tris-HCl buffer, containing mercaptoethanol and EDTA, and centrifuged at 37,000 x g for 30 min. Ammoniu sulfate was added to the supernatant to 20% of saturation. The mixture was stirred at 4°C for 30 min and then centrifuged at 37,000 x g for 10 min. After centrifugation, ammonium sulfate was added to the supernatant to 40% of saturation. The mixture was stirred at 4°C for 30 min and centrifuged at 37,000 x g for 10 min. The precipitate was dissolved in buffer to one-half of the original volume of the crude extract. This entire ammonium sulfate fractionation was then repeated. The second 40% ammonium sulfate precipitate was dissolved in buffer to one-fourth of the original volume of the crude extract. Aliquots of the latter were quick-frozen in liquid nitrogen and stored at -20°C. They retained activity for several months. The aliquots were dialyzed for 1 to 2 hr before use in the guanyl cyclase assay against five hundred volumes of 0.05 M tris-HCl, pH 7.6, containing 10 mM mercaptoethanol. The dialysis tubing was cleaned by boiling in several changes of distilled water before use.

B. Guanyl Cyclase Assay

1. Reaction Mixture

The final volume of the reaction mixture is 75 μl. We use four reaction mixture solutions:

(1) a buffer mix containing caffeine, $MnCl_2$, and mercaptoethanol (25 μl); (2) a GTP regenerating system which also contains bovine plasma albumin (5 μl); (3) the [α -^{32}P]-labeled GTP substrate solution (9 mM, 10 μl); (4) a solution of 20 mM cyclic GMP to act as a "catcher" (10 μl) to minimize destruction of [^{32}P]-cyclic GMP by phosphodiesterase. The remainder of the 75 μl consists of 25 μl of the enzyme solution to be assayed (or of a smaller amount of enzyme, with water making up the difference). The final concentrations of the reaction mixture constituents are as follows: tris-HCl, pH 7.6, 50 mM; caffeine, 20 mM; $MnCl_2$, 6 mM; mercaptoethanol, 10 mM; cyclic GMP, 2.67 mM; (α -^{32}P) GTP, 1.2 mM; creatine phosphate, 5 mM; creatine phosphokinase (75 units per mg), 15 μg; and bovine plasma albumin, 100 μg.

2. Reaction Mixture Solutions

a. Buffer Mix. This is made up shortly before the assay from three stock solutions, the first containing tris-HCl buffer and caffeine, the second, $MnCl_2$, and the third, mercaptoethanol.

A stock solution of 0.167 M tris-HCl and 0.067 M caffeine is prepared by dissolving 2.019 g of tris (free base) and 1.293 g of caffeine in distilled water. The pH is adjusted to 7.6 by the addition of HCl. The mixture is diluted to a final volume of 100 ml. It may be stored at room temperature. However, to cut down on bacterial growth, we have kept most of it in the refrigerator and at intervals removed small amounts to room temperature (after redissolving the caffeine).

The 0.3 M $MnCl_2$ stock solution is prepared by dissolving 5.937 g of $MnCl_2 \cdot 4H_2O$ in 100 ml of distilled water. This is kept at room temperature in a brown bottle. The 0.75 M mercaptoethanol is prepared by diluting 5.13 ml of mercaptoethanol to 100 ml with distilled water. Store in the refrigerator.

At the time of the assay, 0.9 ml of the buffer-caffeine stock solution is pipetted into a test tube. To this is added 0.06 ml of the 0.3 M $MnCl_2$ and 0.04 ml of the 0.75 M mercaptoethanol stock solutions. This buffer mix is kept on ice until used.

b. GTP Regenerating System. This contains creatine phosphate, creatine phosphokinase, and bovine plasma albumin in a single solution. Five milliliters of this solution contains 100 mg of bovine plasma albumin, 15 mg creatine phosphokinase (Sigma, 75 units per mg), and 375 μmoles of creatine phosphate. The hydration state of the creatine phosphate (disodium salt) will vary. This stock solution is dispensed in 0.5-ml amounts in 10 x 75 tubes, quick frozen, and stored in the freezer.

c. [α -^{32}P]-GTP. The specific activity of [α -^{32}P]-GTP now commercially available is very high, on the order of 10 Ci/mmole. We use about 500,000 cpm of the radioactive GTP in each reaction mixture. Assuming 100% counting efficiency and a specific activity of 10 Ci per mmole, this amounts to 0.0225 nmoles of GTP. This may be ignored in the calculation of substrate specific activity, since we have 90 nmoles of cold GTP in the reaction mixture.

We have found that the blank radioactivity in our guanyl cyclase assay [10] is due to certain unknown radioactive contaminants of the [α -^{32}P]-GTP. We also found that the blank increases during storage of the undiluted carrier-free material. However, if cold GTP is added, the blank does not rise. The following procedure is designed to provide for this dilution and still allow for the use of a constant specific activity of substrate (notwithstanding the short half-life of ^{32}P).

The carrier-free [α -^{32}P]-GTP is diluted with distilled water until 1 μl contains approximately 1 x 10^6 cpm. This solution is mixed with an equal volume of 18.0 mM unlabeled GTP. The resulting stock solution contains 9.0 mM GTP and 0.5×10^6 cpm per μl. It may be mixed with 9.0 mM cold GTP in increasing proportions as the ^{32}P decays. Ten microliters of the mixture is used in the reaction mixture, resulting in an [α -^{32}P]-GTP concentration of 1.2 mM, and 0.5×10^6 cpm per assay. To achieve this, immediately after preparation of the 9.0 mM [α -^{32}P]-GTP stock solution, one part of this would be mixed with nine parts of 9.0 mM cold GTP. After 14 days it would be necessary to mix the two solutions in equal proportions. After 47.5 days, the radioactive GTP stock solution would be used directly, since it would have decayed to 10% of the original activity.

To make 18.0 mM GTP solution dissolve 21.715 mg disodium GTP·$2H_2O$ (P-L Biochemicals) in distilled water, adjust the pH to 7.6 with sodium hydroxide, and dilute to 2.0 ml.

d. Cyclic GMP Solution. A 40.0 mM stock solution is prepared by dissolving 33.6 mg of cyclic GMP, sodium salt (Boehringer) in distilled water. The pH is adjusted to 7.6 with sodium hydroxide. The solution is diluted to a final volume of 2.0 ml with distilled water and stored in the freezer. This solution contains 0.2 μmole per μl, the amount of cyclic GMP we use in the standard reaction mixture as a "catcher" for [^{32}P]-cyclic GMP formed. Under these conditions we dilute the stock solution with an equal volume of distilled water and dispense 10 μl to each reaction tube. However, when the tissue to be assayed has a high cyclic nucleotide phosphodiesterase activity, e.g., brain, it is necessary to double the amount of cold cyclic GMP; 10 μl of the stock solution are therefore used.

e. Stop Solution. The stop solution contains 0.1 M EDTA and [^{3}H]-cyclic GMP. Enough of the labeled nucleotide is added to 0.1 M EDTA so that 20 μl of the mixture contains about 30,000 cpm.

1. Stock solution of 0.1 M EDTA. Into about 80 ml of distilled water add 2.92 g of ethylenediaminetetraacetic acid (10.0 mmoles), and 3.62 g of tris base (30 meq). Stir with a magnetic mixer until dissolved. The pH of the solution will be near neutral. Add solid tris base, with mixing, until the pH is 7.6.

2. [^{3}H]-cyclic GMP. Tritiated cyclic GMP is available commercially. The material we are currently using has a specific activity of 4.47 Ci per mmole. It is important that this compound be radiochemically pure, or the activity calculations will be in error. This becomes particularly evident when studies are

made of the activity of an enzyme preparation as a function of time. Such a curve should extrapolate back to zero activity at zero time. If it does not, the purity of the [^{3}H]-cyclic GMP should be suspected. Radiochemical decomposition products are generated even when a solution is stored frozen.

The [^{3}H]-cyclic GMP may be purified by several methods. We prefer thin-layer chromatography since the development solvents can often be removed by evaporation and the nucleotide can be eluted in a small volume without the necessity of lyophilization. In order to remove nucleotide contamination we streak the sample on a silicic acid-fiberglass medium, ChromAR 1000 (Mallinckrodt Chemical Works). The sheet is developed with absolute ethanol - conc. NH_4OH (5:2, v/v) [6]. After development the sheet is dried and the [^{3}H]-cyclic GMP visualized with ultraviolet light and marked with a pencil. The marked strip is cut from the sheet and the nucleotide eluted with water. This is best done by hanging the strip from the trough of a small descending chromatography tank. The nucleotide will be washed off the strip in a very small volume of water.

Contamination with guanine and guanosine may be removed by streaking the sample on a cellulose thin-layer chromatography plate [11]. We use 500-μm Avicel plates (Analtech). The plate is developed with water. After drying, the [^{3}H]-cyclic GMP will be found just behind the front (R_f = 0.9) while guanosine and guanine are retarded (R_f = 0.45 and 0.22, respectively). The area of the plate containing the nucleotide is removed by scraping and the [^{3}H]-cyclic GMP extracted from the scrapings with water.

3. Assay Conditions

The reaction is performed in 10 x 75 mm tubes and incubated with shaking at 30°C. We use a New Brunswick water bath shaker with rotary drive, set at 150 rpm. The reaction is initiated by addition of the enzyme. Both the reaction tubes and the enzyme preparation are incubated in the water bath for about 5 min before the reaction is initiated. At the end of the reaction time, generally 5 min, the reaction is stopped by adding 20 μl of 0.1 M EDTA, pH 7.6, containing about 30,000 cpm of [^{3}H]-cyclic GMP. The tube is immediately placed in a heating block set at 100°C. After 3 min, it is removed from the heating block and placed in an ice bath.

With the same pipettes previously used for each solution, 10 μl of the 9.0 mM (α -^{32}P) GTP and 20 μl of the stop solution (containing [^{3}H]-cyclic GMP) are placed into separate scintillation vials. Prepare 3 vials with each solution. To each vial is added 2.0 ml of column buffer and 10.0 ml of Bray's scintillation solution [9]. These vials are necessary in order to calculate enzyme activity.

4. Purification of the Reaction Product

The purification of the ^{32}P-labeled-cyclic GMP has been a difficult problem, and the methods for accomplishing this have undergone considerable change in our laboratory. We have used three methods with the last one developed now in routine use.

a. Method A. This is the procedure described by White and Aurbach [6] and is still the method that consistently yields the lowest blanks. The guanyl cyclase reaction is stopped by the addition of either 10 μl of 40 mM GTP containing approximately 30,000 cpm of [^{3}H]-cyclic GMP, followed by heating for 3 min at 100°C. The samples are removed to an ice bath until all assays are completed. Then 1.0 ml of 0.05 M HCl is added to each sample and the samples are mixed vigorously on a Vortex mixer and centrifuged for 5 to 10 min in a clinical centrifuge. The supernatant from each sample is applied to a 0.6 x 10 cm column of Dowex AG 50W-X8 (100-200 mesh), previously equilibrated with 0.05 M HCl. When the supernatant has entered the bed, 1.0 ml of 0.05 M HCl is added. When this has entered the bed, 5.0 ml of 0.05 M HCl is added. These column effluants are discarded. Cyclic GMP is eluted with a second 5.0 ml of 0.05 M HCl and the eluate is collected in a 16 x 100 mm test tube. Recovery of the [^{3}H]-cyclic GMP in this fractic is approximately 70 to 80%. The fractions are evaporated overnight to dryness at 40°C in the collection tubes, with a current of dry air.

The residue in each collection tube is dissolved in 75 μl of 50% ethanol. By means of a capillary tube the solution is transferred and applied 1 in from the end of an 8-in sheet of ChromAR Sheet 1000. The spot are dried, and ascending chromatography is carried out with the solvent system of 99% ethanol:conc. ammonium hydroxide (5:2, v/v), for approximately 2 hr. This time exceeds the time required for the solvent front to reach the end of the sheet but is necessary for adequate radiochemical purification. The areas corresponding to cyclic GMP are visualized with ultraviolet light and marked with a pencil while the sheets are wet. The sheets are dried and the marked spots containing cyclic GMP are cut out and placed in scintillation vials. Following the addition of 2 ml of water and 10 ml of Bray's scintillator solution [9], the mixture is shaken vigorously in the capped vials to disperse the ChromAR sheet. Total recovery of the [^{3}H]-cyclic GMP is approximately 40 to 50%.

b. Method B. A less time-consuming method of purification is by chromatography on pretreated ChromAR sheet 1000. The use of pretreated ChromAR sheets permits omission of the column chromatography and evaporation steps in Method A. Pretreatment of the sheets is achieved by dipping them in a solution of 5% (w/v) boric acid in a mixture of 99% ethanol:conc. HCl:H_2O (95:1:4), and drying overnight in the hood.

The guanyl cyclase reaction is stopped as described in Method A. Through the use of a capillary tube, each sample is transferred to pretreated ChromAR sheet and chromatographed, eluted, and counted as was described in Method A. Total recovery of the [^{3}H]-cyclic GMP by Method B is 60 to 70%.

While this method was satisfactory for certain lots of [α -^{32}P]-GTP, it gave unacceptably high blanks with others. It was abandoned after the purification procedure on alumina columns was developed.

c. Method C. A third method of product purification,

developed by White and Zenser [10], is by column chromatography on neutral alumina. In this method, the column effluants are collected directly in the scintillation vials and prepared for radioassay.

After all of the reaction mixtures have been incubated with enzyme and the reactions stopped, 1.0 ml of 0.05 M tris-HCl buffer, pH 7.6, is added to each sample. The tubes are mixed on a Vortex mixer and centrifuged for 5 to 10 min in a clinical centrifuge. Each supernatant is decanted onto a column containing approximately 1.0 g of neutral aluminum oxide which has been equilibrated with the 0.05 M tris-HCl buffer, pH 7.6. We use glass columns, made from 9-mm o.d. standard wall glass tubing, 15 cm long, with one end drawn down to an opening of about 3.5 mm. The other end is joined to a 65-mm length of 25-mm o.d. tubing, which acts as a reservoir. The column tip is packed with a 1-in square of household paper toweling (Teri, Kimberly-Clark Corp.). The columns are supported by Plexiglass racks, each of which holds 40 columns. The rack mounts above either an open Plexiglass box, or another rack constructed to hold scintillation vials, each column draining directly into a vial. Rows of thermometer clamps have also been satisfactory.

The reaction tubes, which have been allowed to rest inverted in the column reservoirs to drain, are removed and discarded. After the samples have entered the alumina, 1.0 ml of the tris-HCl buffer is added to each column with an automatic pipetting syringe. Both the sample applied and this milliliter of buffer are allowed to wash through the alumina and are collected in the box. The rack of columns is then lifted from above the box and placed onto the rack of scintillation vials. Another 2 ml of tris-HCl buffer is applied to each column and allowed to drain into the scintillation vials. This fraction contains about 60% of the cyclic GMP applied to the column. Ten milliliters of Bray's scintillation mixture is added to each vial and the radioactivity measured in a liquid scintillation spectrometer.

With certain batches of alumina, most of the cyclic GMP will be eluted in the fourth and fifth rather than the third and fourth milliliters. The peak elution volume must be determined for each batch. In this case, the application volume is followed by 2 ml of wash buffer. After these are discarded, cyclic GMP is eluted in another 2 ml of buffer.

Most of our experience has been with the neutral aluminum oxide supplied by Sigma Chemical Company.

A sample of neutral alumina from E. Merck (Darmstadt) gave about 5% higher [^3H]-cyclic GMP recoveries than the Sigma alumina in current use. However, the calculated ^{32}P blank was also slightly higher.

5. Calculations

The specific activity of guanyl cyclase is expressed as picomoles of cyclic GMP formed per minute per milligram of protein when assays are performed at temperatures other than 30°C. At 30°C, specific activity is expressed in microunits (μU) per milligram protein in accord with the I.U.B. convention (1965). One microunit is defined as the amount of enzyme catalyzing the transformation of 1 pmole of substrate per minute at 30°C.

The calculations used to determine the picomoles of cyclic GMP formed during the guanyl cyclase reactio are as follows. The sample radioactivity is first corrected for the recovery of [^3H]-cyclic GMP during the purification and for the efficiency of the method of purification (^{32}P activity in the blank). By this method, corrected sample radioactivity C would be:

$$C = (\text{cpm } ^{32}P_a)\left(\frac{\text{cpm } ^3H_s}{\text{cpm } ^3H_a}\right) - (\text{cpm } ^{32}P_b)\left(\frac{\text{cpm } ^3H_s}{\text{cpm } ^3H_b}\right)$$

where the subscript "a" refers to the product of the

cyclase reaction, "s" refers to the labeled compound added to the reaction mixture, and "b" refers to the blank. The specific activity (A) of the [α -^{32}P]-GTP added to the reaction mixture in cpm per pmole would be:

$$A = \frac{\text{cpm}[\alpha\text{-}^{32}\text{P}]\text{-GTP}_S}{\text{pmoles GTP} + \dfrac{\text{cpm}[\alpha\text{-}^{32}\text{P}]\text{-GTP}_S}{(\text{pCi/pmole}[\alpha\text{-}^{32}\text{P}]\text{-GTP})(2220\ \text{cpm/pCi})}}$$

As we have previously mentioned, with the high specific activity [α -^{32}P]-GTP now available, the amount of GTP contributed by the labeled nucleotide becomes negligible and A simplifies to:

$$\frac{\text{cpm}\ [\alpha\ \text{-}^{32}\text{P}]\text{-GTP}}{\text{pmoles GTP}}$$

The number of picomoles of cyclic GMP formed during the reaction is equal to C/A.

C. Protein Determination

We have expressed enzyme activity in terms of the protein content of the preparation. This was particularly necessary in the development of the assay, when we found that the amount of homogenate supernatant we assayed affected the enzyme activity measured (Section II). We determine the protein content of the sample aliquot with the same pipet used to add the enzyme. We have used two methods of protein determination, the microbuiret procedure of Goa [12] and the Folin phenol procedure of Lowry et al. [13]. The first has a wider range and was useful in the studies on the effect of sample size. Reducing agents interfere with both; therefore, when the homogenizing medium contains mercaptoethanol or dithioerythritol, the sample protein is first precipitated. We use 2.0 ml of silicotungstic acid [14] since this precipitate is more easily redissolved then that obtained with trichloroacetic acid. After standing 30 min, the samples are centrifuged for 10 min in a clinical centrifuge. The supernatants are drained from the tubes. In the microbiuret procedure, the precipitate is dissolved in 2.0 ml of 3% sodium hydroxide. Fifteen minutes after the addition of 0.1 ml of Benedict's reagent, the extinction is read at 330 nm. In the procedure of Lowry et al., the precipitate is dissolved in 4.0 ml of the copper carbonate-sodium hydroxide reagent. The tubes are placed in a water bath at 37°C for 10 min. Then to each tube is added 0.4 ml of 1 N phenol reagent and the tubes are again placed in the water bath for 20 min, after which they are read at 725 nm. Bovine plasma albumin is used as standard.

IV. PURIFICATION OF GUANYL CYCLASE

Both Hardman and Sutherland [5] and ourselves [6] have shown that ammonium sulfate fractionation of supernatant fractions of lung and other tissues resulted in increased specific activity of guanyl cyclase; a four-fold increase was obtained with extracts from rat or beef lung. Hardman and Sutherland [5] were able to increase specific activity another three-fold by lowering the pH to 5.1 and then to 4.5 and collecting the precipitates. With beef lung we have not been able to redissolve the precipitates resulting from acidification of tissue extracts, either before or after ammonium sulfate. The successful use of a pH step clearly depends upon the tissue processed, since with rat liver Hardman and Sutherland used such a step before ammonium sulfate precipitation and increased the final specific activity 20-fold.

We have carried the purification of guanyl cyclase from beef lung several steps further, utilizing ion-exchange chromatography, gel filtration, and ultrafiltration. The final specific activity was 445 times that of the original tissue extract. While this preparation is still far from homogenous when examined by disc-gel electrophoresis, it represents the highest purification yet reported [15].

A. Preparation of Starting Material

The beef lungs used were from a local slaughterhouse and were still warm. The lungs were placed in ice and transported to a local meat processor. Here the lungs were minced into small pieces slightly larger than 0.5 cm in diameter. All further steps were carried out at 4°C.

Nine hundred grams of minced lung were placed in a stainless steel Waring blender with 2700 ml of 0.05 M tris-HCl buffer, pH 7.7, containing 10 mM mercaptoethanol and 0.5 mM EDTA (TEM buffer). This mixture was homogenized at the "Hi" setting for 1 min, stopped for 1 min, and then homogenized for 1/2 min more at the "Hi" setting. The homogenate was centrifuged at 14,000 x g for 30 min. The pellet was discarded and supernatant filtered by gravity through glass wool and saved. This homogenization was repeated until 6600 ml of supernatant were obtained.

B. Ammonium Sulfate Fractionation

Solid ammonium sulfate was slowly added to the supernatant with rapid stirring by a magnetic stirrer until the solution was 20% of saturation. Care was taken that no local concentrations of ammonium sulfate existed during the addition. The mixture was stirred for 30 min, and then it was centrifuged at 14,000 x g for 30 min. The pellet was discarded. Solid ammonium sulfate was again added to this supernatant until 40% of saturation was reached. The stirring was continued for 30 min again, while the pH was maintained above pH 7.0.

The pellet from the 40% ammonium sulfate fractionation was resuspended in 3300 ml of the starting buffer and stirred for 1 hr to insure complete solubilization of the pellet. To this solution solid ammonium sulfate was again added as described above, i.e., first to 20% of saturation and then to 40% of saturation. This second 40% pellet was resuspended in 1650 ml of starting buffer and stirred for 1 hr.

The final resuspended solution was dialyzed for 12 hours against 16 liters of buffer, then for 12 hr against 32 liters of buffer, and finally for 8 hr against another 32 liters of buffer. The dialyzed preparation was then quick frozen with liquid nitrogen and stored at -20°C. The results of ammonium sulfate fractionation are shown in Table 2. The specific activity and recovery of cyclase in this particular experiment were extremely high. Usually there is a four-fold purification at this step with a 50% recovery. The only procedural changes for this experiment were

TABLE 2

Purification of Bovine Lung Guanyl Cyclase by Ammonium Sulfate Fractionation[a]

Fraction	Volume (ml)	Protein (g)	Specific activity: Cyclase (μU/mg)	Specific activity: PDE (mU/mg)	Percent recovery: Cyclase	Percent recovery: PDE
I. 14,000 x g supernatant	6600	165	20.8	2.5	100	100
II. 0-20% $(NH_4)_2SO_4$ #1						
Supernatant	6800	122	18	1.9	96	56
Precipitate	820	9	14	-	2	-
III. 20-40% $(NH_4)_2SO_4$ #1						
Supernatant	6800	75	0	0.6	0	11
Precipitate	3300	45	62	5.0	75	54
IV. 0-20% $(NH_4)_2SO_4$ #2						
Supernatant	3300	38.6	91	5.1	102	48
Precipitate	3300	10.6	31	2.0	10	5
V. 20-40% $(NH_4)_2SO_4$ #2						
Supernatant	3150	7.9	0	0	0	0
Precipitate	1650	22	168	9.0	108	48

[a]All fractions were desalted on a Sephadex G-25 column before assay. Protein was determined by the Inchiosa method [17]. One unit of phosphodiesterase activity represents 1.0 μmole of cyclic GMP utilized per minute at 30°C.

the mincing of lungs into small pieces at a commercial meat processor and the use of a 1-gal stainless steel Waring blender. These two changes may have caused a larger yield of cyclase. The final ammonium sulfate fraction seemed to be unstable to storage when frozen. However, if this fraction was immediately dialyzed and then quick frozen, it was quite stable when stored at -20°C.

As has been previously reported [6], guanyl cyclase precipitates between 20 and 40% of ammonium sulfate saturation. In this experiment we also determined cyclic nucleotide phosphodiesterase activity, using a modification of the method of Thompson and Appleman [16, see also Chapter 6]. The phosphodiesterase reaction mixture contained 5.3×10^{-5} M [^{3}H]-cyclic GMP, 0.05 M tris-HCl, pH 7.6, 5 mM $MgCl_2$, 1 mM dithioerythritol, 100 μg bovine plasma albumin, 10 μg *Crotalus atrox* venom, and 5 μl of the enzyme preparation in a final volume of 75 μl. The reaction was initiated by addition of the substrate and, after 10 min incubation at 30°C, was stopped by adding 1.0 ml of a 1:3 slurry of Dowex 1-X8 (200-400 mesh). As shown in Table 1, the phosphodiesterase activity accompanied the cyclase activity. There was an eight-fold purification of cyclase and a four-fold purification of phosphodiesterase.

C. Effect of Prolonged Dialysis

We have found that ammonium sulfate is an inhibi of guanyl cyclase [8], so the ammonium sulfate fracti had to be desalted before assay. This was accomplish with a small Sephadex G-25 column. However, when an ammonium sulfate preparation is dialyzed against TEM buffer for 30 hr with two changes of buffer, a doubli of specific activity resulted (Table 3). Approximate 20% of the protein precipitated during the dialysis, we considered the possibility that an inhibitor might have been removed. However, when the precipitate was resuspended in 0.2 M NaCl and an aliquot added to the assay mixture, no inhibition of activity resulted.

D. DEAE - Cellulose Chromatography

The dialyzed ammonium sulfate fraction was appli to a DEAE - cellulose column (Figure 6). Guanyl cycl binds to the column under these conditions, while a large peak of protein and nucleic acid passes through After washing the column to remove this unbound mater a linear gradient of NaCl is started. Enzyme activit begins to elute at apporoximately 0.08 M NaCl. The t containing the greatest activity are pooled (185 ml). As can be seen from Table 3, this 185 ml contained on 163 mg of protein, as compared to the 966 mg applied

TABLE 3

Purification of Bovine Lung Guanyl Cyclase [a]

Fraction	Volume (ml)	Protein (mg)	Total Units (mU)	Specific activity (μU/mg)	Recovery (%)
I. 14,000 x g supernatant	368	8280	105.0	13	100
II. Ammonium sulfate	92	1196	56.3	47	54
III. Dialyzed ammonium sulfate	92	966	116.0	119	110
IV. DEAE-cellulose	185	163	148.0	910	141
V. Amicon UM20E	13	130	231.0	1776	220
VI. Sephadex G-200					
Peak I	29	67	7.3	109	7
Peak II	39	18	31.0	1719	30
VII. Amicon XM-300					
Peak I	11.5	66.5	6.6	99	6
Peak II	9.8	11.3	65.6	5783	63

[a]Protein was determined by the Warburg and Christian method [18]. Fractions I and II were desalted on a Sephadex G-25 column before assay.

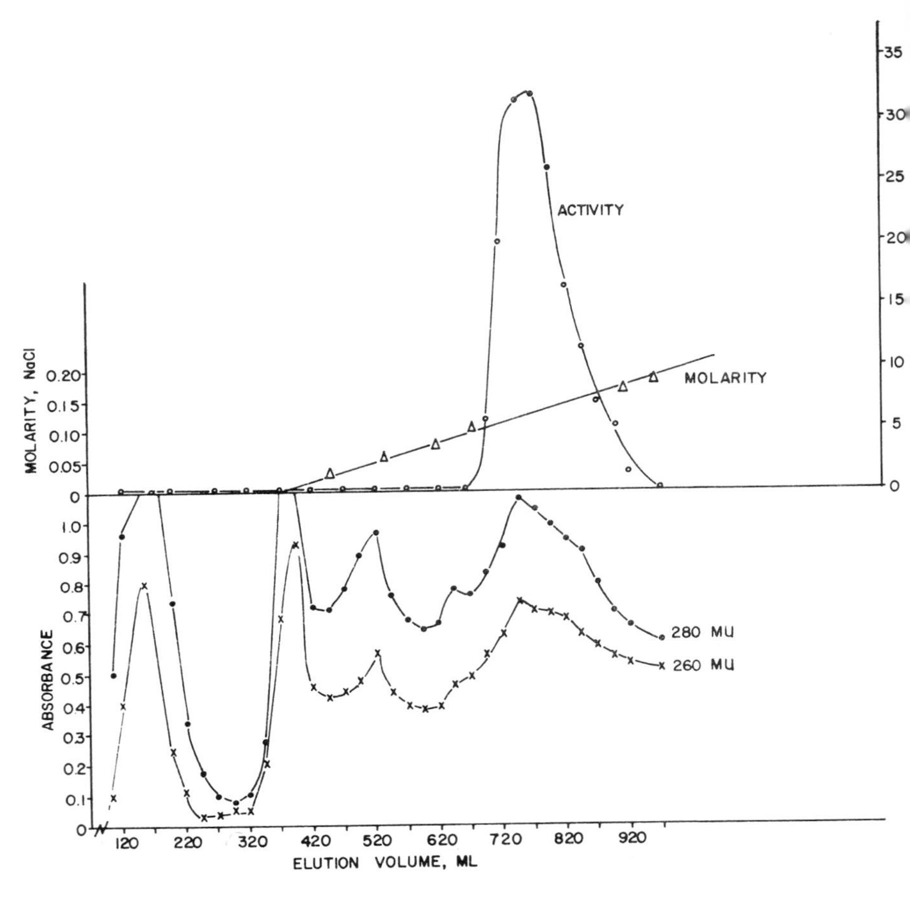

Fig. 6. Elution diagram of guanyl cyclase from a DEAE-cellulose column. The applied enzyme contained 9 mg of protein in 92 ml of TEM buffer. It was adsorbed on a 2.5 x 27 cm column of Whatman DE-52. After apply column volume of TEM buffer, a linear gradient of sodium chloride was applied. There was 850 ml of 0.5 M NaCl in TEM buffer in the reservoir and an equal volume of TEM buffer in the mixing chamber. Flow rate was 45 ml per hour and fractions of 5 ml were collecte

the column. The total units of activity increased as a result of DEAE - cellulose chromatography, going from 116 mU to 148 mU.

E. Amicon UM20E Ultrafiltration

The 185 ml eluted from the DEAE - cellulose column was concentrated to 13 ml with an Amicon UM20E ultrafiltration membrane. This concentration resulted in a doubling of the specific activity although there was only a 33-mg loss in protein (Table 2). At this point we had more than doubled the total units of activity present in the original 14,000 x g supernatant.

F. Gel Filtration

Part of the Amicon UM20E concentrate was applied to a Sephadex G-200 column (Fig. 7). A small peak of activity appeared in the void column while a larger peak was eluted with a K_d of 0.2. It has not yet been established if the activity in the void volume represents a different form of the enzyme or is a complex of the enzyme with some other components of that peak. We have found that if the larger peak of activity is rerun on the same column, the first peak does not reappear. Furthermore, if a dialyzed ammonium sulfate preparation is directly applied to the Sephadex G-200 column, a pattern is obtained that does not differ qualitatively

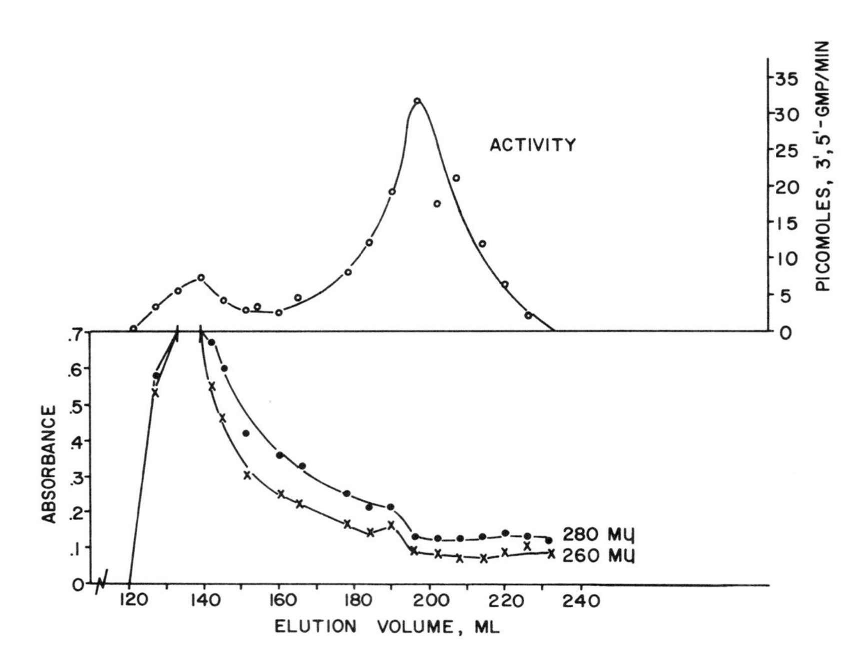

Fig. 7. Gel filtration of guanyl cyclase on Sephadex G-200. The 185 ml containing the peak activi eluting from the DEAE-cellulose column (Fig. 6) were concentrated to 13 ml with an Amicon UM20E ultrafilter Of this concentrate, 4.5 ml (containing 45 mg of prote were applied to a 2.5 x 83 cm Sephadex G-200 column. The elution buffer was 0.05 M tris-HCl, pH 7.7, contai ing 0.5 mM EDTA and 0.25 mM dithiothreitol. The flow rate was 17 ml per hour and 3 ml fractions were collected.

from Fig. 7 with respect to the elution of enzyme activity. There was much more protein and nucleic acid appearing in the void volume. If the first peak of activity was due to complex formation, it might be expected to represent a larger fraction of the total activity under these conditions, but it does not. These results suggest that there are two separate guanyl cyclase activities eluting from Sephadex G-200, and that they are not readily interconvertible.

G. Amicon XM300 Ultrafiltration

The two peaks obtained from the Sephadex G-200 column were separately pooled and concentrated using an Amicon XM300 membrane. This resulted in doubling of the total activity present in peak II while 37% of the protein was lost. The specific activity was, therefore, increased more than three-fold. The activity of peak I was unchanged by ultrafiltration. The specific activity of peak II (5783 μU per mg) represents a 445-fold purification of the original 14,000 x g supernatant. However, this peak also contains considerable phosphodiesterase activity.

The increase in activity that we obtained when peak II was concentrated by ultrafiltration suggests that the activity of guanyl cyclase may be dependent on enzyme concentration. This could also account for the increase

in activity observed after DEAE - cellulose chromatogr
and ultrafiltration. Peak I showed no increase in activity with ultrafiltration; peak I is also more sensitive to freezing than peak II. Our working hypothesis is that peak I is an aggregated form of peak II.

REFERENCES

[1] D. F. Ashman, R. Lipton, M. M. Melicow, and T. D. Price, Biochem. Biophys. Res. Commun., 11, 330 (1963).

[2] E. W. Sutherland and T. W. Rall, J. Am. Chem. Soc., 79, 3608 (1957).

[3] T. D. Price, D. F. Ashman, and M. M. Melicow, Biochem. Biophys. Acta, 138, 452 (1967).

[4] E. Bohme, K. Munske, and G. Schultz, Nauyn-Schmeidebergs Arch. Pharmakol., 264, 220 (1969).

[5] J. G. Hardman and E. W. Sutherland, J. Biol. Chem., 244, 6363 (1969).

[6] A. A. White and G. D. Aurbach, Biochim. Biophys. Acta, 191, 686 (1969).

[7] T. W. Conway and F. Lipmann, Proc. Nat. Acad. Sci. U.S., 52, 1462 (1964).

[8] S. J. Northup, Ph.D. Thesis, University of Missouri, Columbia, 1971.

[9] G. A. Bray, Anal. Biochem., 1, 279 (1960).

[10] A. A. White and T. V. Zenser, Anal. Biochem., 41, 372 (1971).

[11] U. Klotz, S. Berndt, and K. Stock, Life Sciences, 11, part 2, 7 (1972).

[12] J. Goa, Scand. J. Clin. Lab. Invest., 5, 218 (1953).

[13] O. H. Lowry, N. J. Rosebrough, A. L. Farr, and R. J. Randall, J. Biol. Chem., 193, 265 (1951).

[14] J. B. Martin and D. M. Doty, Anal. Chem., 21, 965 (1949).

[15] T. V. Zenser, Ph.D. Thesis, University of Missouri, Columbia, 1971.

[16] W. J. Thompson and M. M. Appleman, Biochemistry, 10, 311 (1971).

[17] M. A. Inchiosa, J. Lab. Clin. Med., 63, 319 (1964).

[18] O. Warburg and W. Christian, Biochem. Z., 310, 384 (1942).

[19] P. Andrews, Biochem. J., 96, 595 (1965).

Chapter 6

CYCLIC 3',5'-NUCLEOTIDE PHOSPHODIESTERASE

Mark Chasin and Don N. Harris

Department of Biochemical Pharmacology
The Squibb Institute for Medical Research
New Brunswick, New Jersey 08903

Cyclic nucleotide phosphodiesterases, whether relatively specific for cyclic AMP or cyclic GMP or accepting either nucleotide equally, are found in virtually every animal tissue examined, from one-cell organisms to man [1-4]. Although a phosphodiesterase specific for cyclic UMP has been found in a dog heart and rat lipocytes [5,6], cyclic UMP itself has not as yet, been detected.

Highly purified phosphodiesterases have not yet been routinely prepared. When conventional means have been used to purify these enzymes, enzyme activity has usually been greatly reduced or lost [7,8]. Highly purified phosphodiesterase has been prepared from frog erthrocytes by Rosen [2]. Butcher and Sutherland [1] have purified an enzyme isolated from beef heart 100-fold, while Nair [7] has purified an enzyme from dog heart 170-fold. Despite the resistance to purification that seems to be characteristic of these enzymes, in the great majority of cells the enzymes are present in high concentrations. This prevalence has allowed dialyzed tissue homogenates, particulate fractions, and supernatant fractions to be assayed directly for the particular phosphodiesterase of interest. Enzyme preparations of this type are easily prepared, and are used by most of the investigators in this field. In our laboratory, we have prepared enzymes of this type from at least ten different sources. In this report, however, we shall give only a detailed description of the preparation of the soluble enzymes from rat brain and cat heart, since it has been our experience that the procedure is applicable for the preparation of phosphodiesterases from many sources.

I. PREPARATION AND PROPERTIES OF THE ENZYME

A. Preparation of the Enzyme

A 1- to 2-kg mongrel cat or 10 to 12 male Sprague-Dawley rats (150-200 g) are killed by cervical dislocation and decapitation, respectively. The cat heart and rat brains are removed and immediately placed on cracked ice; the auricular tissue of the heart is cut away. The following steps are all carried out at 0-4°C. The organs are minced and homogenized in 5 to 10 vol of 0.05 M imidazole buffer (pH 7.5), which also contains 5 mM dithiothreitol (Cleland's reagent) in the heart preparations. The use of either tris or cacodylate buffer results in an active preparation containing half of the specific activity of enzyme prepared in imidazole, in agreement with the demonstrated ability of imidazole to stimulate the enzyme [1]. We have found that phosphat buffer markedly inhibits the enzyme and, therefore, should not be used. (A portion of the phosphodiestera activity is particulate, and can be converted to a soluble form in brain homogenates, at least, by sonication at this point, if desired). The homogenates are immediately centrifuged for 15 to 20 min at 39,000 x g. The supernatant fractions are adjusted to 50% saturation by very slow addition of an equal volume of neutral, saturated ammonium sulfate solution the pH is adjusted to 7.5 with 1 N NaOH, and the mixture is allowed to stand for 1 hr at 0°C. The solutions are centrifuged as before, and the precipitates are taken up in a minimum volume of the imidazole buffer and dialyzed against 20 vol of the same buffer overnight. Protein concentrations were from 5 to 7 mg/ml for the cat heart preparation, by a micromodification [9] of the biuret method [10], and 15 to 20 mg/ml for rat brain phosphodiesterase, as determined by the Lowry procedure [11].

This procedure also has been used successfully in our laboratories to prepare phosphodiesterase from dog blood platelets, rat adrenal glands, fetal-calf pancreas, guinea pig lung, rat heart, isolated rat lipocytes, and rabbit brain. Enzyme prepared from slime mold required 90% ammonium sulfate rather than 50%, but was otherwise prepared by the above procedure. Fractions obtained from the initial 39,000 x g precipitate may be washed thoroughly with buffer and assayed directly for phosphodiesterase activity present in particulate cellular fractions, if desired.

B. Storage of Phosphodiesterase

1. Frozen Preparations

After determination of the specific activity of the cat heart preparation from a plot of activity against protein concentration, suitable enzyme dilutions are prepared for storage as follows. The dilution medium consists of 60 mM tris buffer that has been adjusted to pH 8.0 with HCl. Human (or bovine) serum albumin is added to the buffer to give a final concentration of 1 mg/ml. Six ml of this diluting medium is pipetted into a plastic vial. To the vial is added a suitable amount of concentrated enzyme so that 50 μl of the diluted enzyme will hydrolyze 20-40% of the cyclic AMP in the assay medium in 10 min. The enzyme preparations in covered vials are immediately frozen at -70°C, and stored at that temperature or at -20°C. Phosphodiesterase preparations stored in this manner maintain their activity for more than nine months.

2. Refrigerator Storage

Unfrozen enzyme from cat heart also may be used if it is stored at 0°-4°C but it loses as much as 40% of its activity in one month under these conditions.

The rat brain enzyme is more stable, and can be stored at 0°-4°C for at least six months with no loss of activity. The stability of other preparations varies widely. Enzymes prepared from either rat lipocytes or adrenal glands are virtually inactive after one week of storage at 0°-4°C, whereas enzyme prepared from rabbit brain has been stored under these conditions for up to 20 months with no significant loss of activity.

C. Properties of the Enzyme

In most tissues, cyclic nucleotide phosphodiesterase is found in both soluble and particulate forms [1,6,12]. As is the case with most particulate enzymes, little is known about the properties of particulate phosphodiesterase, so the rest of this discussion will be concerned with the soluble form.

The pH optimum for activity is rather broad, with

a peak at about pH 8.0 [1-4,7]. We have found that below pH 5.5, the enzyme from rat brain precipitates. At pH 5.0, this precipitation is virtually complete, but upon neutralization the precipitated protein redissolves and is fully active. Below pH 5.0, irreversible denaturation of the enzyme occurs. The enzyme survives 1 hr at 0°C at pH values up to 10, with little or no loss of a tivity.

The activity of the phosphodiesterase from rat brain shows a sharp drop after exposure to temperatures above 45°C [13]. The enzymes from various sources require divalent metal ions, usually Mg^{2+}, for maximum activity [1,2,7,13].

The enzymes from most mammalian sources are inhibited by methylxanthines, and are stimulated by imidazole [1]. A naturally occurring protein activator has also been reported [8]. A large number of inhibitors of this enzyme has been reported, the most potent of these being 4-(3-butoxy-4-methoxybenzyl)-2-imidazolidinone (Ro 20-1724) for the enzyme prepared from rat erythrocytes (50% inhibition at about 10^{-7} M) [14], 1-ethyl-4-(isopropylidenehydrazine)-1H-pyrazole-(3,4-b)-pyridine-5-carboxylic acid ethyl ester hydrochloride (SQ 20,009) for the enzyme prepared from rat or rabbit brain (K_i of 2 x 10^{-6} M for the rat brain enzyme) [15], and papaverine for the enzyme prepared from coronary arteries (K_i of 5.6 x 10^{-6}M) [16]. It should be noted that Ro 20-1724 is a far more potent inhibitor of rat erythrocyte phosphodiesterase than of the enzyme prepared from cerebral cortex [17], whereas SQ 20,009 is a relatively specific inhibitor for enzyme prepared from brain [15]. These findings suggest that phosphodiesterase prepared from different tissues differ considerably, at least with respect to inhibition by different agents, and allows for the possibility of designing tissue-specific phosphodiesterase inhibitors.

Several investigators have reported that multiple forms of cyclic nucleotide phosphodiesterase exist in various tissues, based on the demonstration of multiple K_m values for the nucleotide [15,18-20], or on separation of several forms by molecular sieve chromatography [20] or electrophoresis [2,21]. The two K_m values for the rat brain enzyme are 4 μM and 120 μM.

II. ASSAY OF PHOSPHODIESTERASE ACTIVITY BY A MODIFICATION OF THE RADIODISPLACEMENT ASSAY

This procedure is a modification of the radiodisplacement assay of Brooker et al. [18]. The procedure may be diagramatically represented as follows:

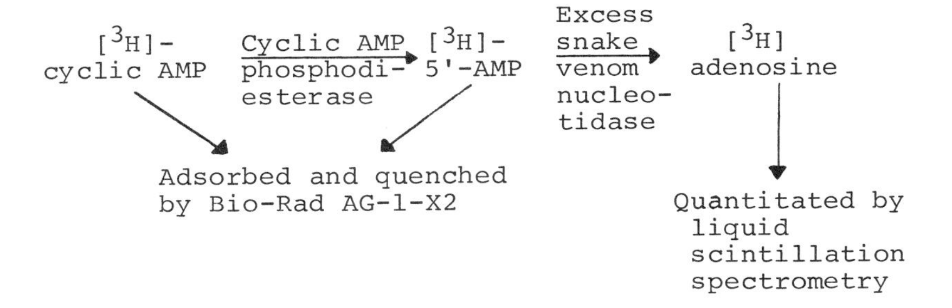

The entire assay is conducted in a plastic scintillation vial, by the addition of the following solutions, in order. (Plastic vials are used because their hydrophobic character causes the solution in the vial to remain as a single drop.) A 50-μl aliquot of 0.05 M tris buffer, pH 8.0, is placed in the vial. If compounds are to be tested for potential inhibition of the enzymatic activity, then a 50-μl aliquot of the inhibitor solution is added to the vial in place of this buffer. Note that the final assay concentration of inhibitor will be three-fold as dilute as that added. All inhibitor solutions are freshly prepared before each assay. We find that the addition of 50-μl of either buffer or test solutions is considerably simplified by the use of either a Biopipetter (Baltimore Biological Laboratories, #05-719) or an Eppendorf Micropipetter (Brinkmann Instruments, Inc.

#22-34-15).

A 50-μl aliquot of the substrate solution containing 120 mM tris-HCl (pH 8.0), 2.5 mM EDTA, 120 mM $MgCl_2$, 120 μM 5'-AMP (this is optional [18], not required), and 0.16 μM [^{3}H]-cyclic AMP is added directly to the drop of buffer or inhibitor already in the vial. The substrate solution is stored at 0°-4°C until used. We use about 220,000 dpm per assay of [^{3}H]-cyclic AMP, with a specific activity of 12.7 Ci/mmole. Cyclic GMP phosphodiesterase may be easily assayed by use of [^{3}H]-cyclic GMP as substrate in this assay; all other components are added as for determination of cyclic AMP phosphodiesterase activity.

A vial containing the stock cat heart phosphodiesterase and human serum albumin (1 mg/ml) in 6 ml of 60 mM tris-HCl (pH 8) is removed from the freezer and allowed to thaw at room temperature. Five ml of the enzyme solution are added to a glass vial containing 5 mg of king cobra venom (*Ophiophagus hanna*). The reaction is started by the addition of 50 μl of this enzyme-venom solution directly to the drop in the vial, which already contains the inhibitor (or buffer) and substrate solutions.

For unfrozen preparations, both cat heart and

rat brain enzyme solutions are prepared in the same manner. To start the reaction, add directly to the drop in the vial, containing inhibitor (or buffer) and substrate solutions, 50 μl of an enzyme mixture that is prepared before each assay and contains human serum albumin (1 mg/ml), lyophilized king cobra venom (Ophiophagus hanna, 1 mg/ml), 60 mM tris HCl (pH 8.0), and a predetermined amount of the phosphodiesterase preparation. The amount of phosphodiesterase preparation added, as determined from a plot of protein concentration against activity, is sufficient to convert 20-40% of the radioactive substrate during the 10-min incubation period. The additions of both the substrate solution and enzyme solution are immensely simplified by use of a Hamilton repeating dispenser (Model PB600) fitted with a Hamilton gas-tight syringe (Model 1002SN). This instrument allows one to deliver 50 consecutive 50-μl aliquots by sequentially pushing a button, and greatly speeds the assay while increasing reproducibility. The uncapped vials are incubated for 10 min at 37°C with shaking at position 4 (about 160 rpm) in a New Brunswick Scientific Model G76 Gyrotory Water Bath. The reaction is stopped by the addition of 0.5 ml of an aqueous slurry (50% settled volume) of Bio-Rad

AG-1-X2, -400 mesh anion-exchange resin. This resin adsorbs unreacted cyclic AMP and 5'-AMP, thereby quenching the scintillation owing to radioactivity of these compounds. (Adenosine remains in solution, and is finally counted by scintillation spectrometry, as described below). The vials are shaken for an additional 30 sec at position 6 (about 240 rpm) of the water bath, at 37°C. The vials are then removed from the water bath and allowed to equilibrate for 10 min at room temperature. Then 10 ml of a scintillation fluid are added (prepared as follows: 8 g of PPO (2,5-diphenyloxazole), 0.6 g of dimethyl POPOP (1,4-bis-[2-(4-methyl-5-phenyloxazolyl)]-benzene), 150 g of naphthalene, 100 ml of 2-ethoxyethan 20 ml of ethylene glycol, and sufficient 1,4-dioxane to bring the final volume to 1 liter. This scintillation cocktail, originally described in the Packard catalog, gives the highest efficiency in this system. Other dioxane-based scintillation mixtures may be used, but will generally give lower efficiencies). After equilibration of the samples at 5°C for 20 to 60 min, count each sample for 1 min in a liquid scintillation counter with a sample compartment cooled to 5°C. If an ambient-temperature counter is used, the cooling step is eliminated. The following controls (three counting vials each) are run with each assay:

enzyme control- 50 μl of the substrate mixture, 50 μl of the enzyme mixture, and 50 μl of 60 mM tris-HCl buffer, pH 8.0;

blank-50 μl of the substrate mixture and 100 μl of 60 mM tris-HCl buffer, pH 8.0

For compounds being tested for potential inhibition of phosphodiesterase activity, percent inhibition is determined according to the following formula:

$$100 - 100 \times \frac{\text{test sample counts} - \text{mean blank counts}}{\text{mean enzyme control counts} - \text{mean blank counts}} = \%\ \text{inhibition.}$$

It has been shown [8] that partial proteolytic digestion of certain purified phosphodiesterases results in an apparent activation of enzymatic activity. For this reason, the procedure described in this section should be used with caution, since phosphodiesterase is incubated along with snake venom, containing proteolytic activity. *Ophiophagus hanna* venom is used because it has only a small amount of proteolytic activity. However, if one is investigating a new phosphodiesterase or purified fractions of this enzyme, it may be important to determine the activity of phosphodiesterase in the absence of venom. A two-step modification of this assay has been developed [19] for this purpose.

$[^3H]$-cyclic AMP (200,000 cpm), $MgCl_2$ (5 mM), buffer (as described for the one-step procedure), and mercaptoethanol (4 mM, optional) are incubated with phosphodiesterase for 10 min at 37°C in a final volume of 0.4 ml. The reaction is stoppped by transfering the vials to a boiling water bath for 3 min. After cooling, 0.1 ml of _Ophiophagus hannah_ venom (1 mg/ml) is added to each vial, followed by incubation for 10 min at 37°C. This reaction is stopped by the addition of the Dowex slurry described for the one-step procedure. After equilibration and addition of scintillation fluid, the samples can be counted.

It has been our experience that using the rat brain and cat heart enzymes described in Section I.A, no differences are observed in the activities of the enzymes as measured by either the one- or two-step procedure. More purified preparations may be expected to show lower activities in the two-step procedure [8].

Whereas the modified radiodisplacement assay is

is generally useful for a wide range of conditions of pH, temperature, substrate concentrations, and potential inhibitor studies, other assays have proven useful either for specific purposes, described in the next section, or in laboratories where equipment or expertise dictate the use of other methods.

III. SPECTROPHOTOMETRIC ASSAYS

These assays, while useful for the specific applications described for each, lack the general usefulness of the modified radiodisplacement assay. Furthermore, their useful range of substrate concentrations is not nearly as large as that of the modified radiodisplacement assay.

A. Coupled Assay with Adenylic Acid Deaminase

This assay is based upon a method first described by Drummond and Perrott-Yee [22]. The procedure measures the decrease in absorbance at 265 nm as adenosine-5'-monophosphate is converted to the corresponding inosine derivative, as shown:

$$\text{Cyclic AMP} \xrightarrow[\text{phosphodiesterase}]{\text{Cyclic AMP}} \text{5'-AMP} \xrightarrow[\text{acid deaminase}]{\text{Excess adenylic}} \text{5'-IMP}$$

The original procedure used adenylic acid deaminase,

purified through six steps by the procedure of Lee [23]
We use either commercially available adenylic acid deaminase (from Sigma), or the elegant one-step purification from rabbit muscle homogenate to homogeneous, crystalline enzyme described by Smiley et al. [24]. The following procedure uses the Smiley et al. preparation of adenylic acid deaminase.

To a semi-micro quartz cuvette (0.5 ml capacity) with a 1-cm light path, contained in a thermostatted cuvette holder of a Gilford (or Beckman model DU) spectrophotometer at 25°C, is added sequentially 0.475 ml. of 0.20 M buffer, pH 7.00, containing 1mM $MgCl_2$; 10 μl of a solution of 9.9 mg of cyclic AMP per 12 ml of buffer (final assay concentration, 50 μM; 5 μl of a solution of 3 mg/ml of adenylic acid deaminase, and, to initiate the enzymatic reaction, 10 μl of a suitable dilution of cyclic AMP phosphodiesterase. We have used all three buffers described in Section II of this chapter, tris-HCl, imidazole, and cacodylate, with about equivalent results, although the specific activity of enzyme prepared in imidazole buffer is, as described previously, about twice that of enzyme prepared in either of the other two buffers. To simplify the additions, all components are added to the buffer in the cuvette with Eppendorf automatic microliter pipettes, available from Brinkmann. These components are mixed, and the decrease in absorbancy at 265 nm is followed with time. At a rate of change of absorbance of 0.01 A/min, a 50-fold excess of adenylic acid deaminase over phosphodiesterase is present. This excess can be checked by including 5'-AMP, rather than cyclic AMP, in the reaction mixture, thereby determining the activity of the deaminase.

Calculations of specific activity are as follows. The average initial absorbance change per minute is obtained from the recording spectrophotometer. This number is divided by 8.86, the factor obtained from the difference in molar absorbancy of AMP and IMP at 265 nm; this calculation yields the activity of the phosphodiesterase preparation in μmoles/min. Dividing further by the phosphodiesterase protein added to the assay yields specific activity, in μmoles/min/mg of protein. Typical values obtained for ammonium sulfate-fractionated rabbit brain preparations, prepared exactly as described in Section I, were about 0.010 μmoles/min/mg of protein in our laboratories. Drummond and Perrott-Yee [22] using a similar preparation, obtained specific activities for rabbit brain preparations of 0.018 μmoles/min/mg of protein.

We have found this assay to be of little value in determining the effect of various compounds on phosphodiesterase activity, since the assay is conducted at 265 nm, and a large proportion of organic molecules absorb quite strongly in this spectral region. It has proven however, to be extremely valuable for a different application.

When phosphodiesterase activity is eluted from columns, or fractions are collected from sucrose density-gradient centrifugations, one generally obtains a large number of fractions, many of which are inactive. This assay can be conveniently used for the preliminary estimation of activity of such fractions. An assay cuvette is prepared as previously described, and 10 μl of the first fraction is added. If no activity results, 10 μl of the second fraction to be assayed is added to the same cuvette. Again, if no activity results, the third fraction is added, and so on. When an active fraction is found, the next fraction is the first to be assayed in a new cuvette. This allows a very rapid survey of all the fractions. Detailed, careful assays are then required only on those fractions indicated by this first assay to contain activity. One should always be aware of potential artifacts introduced by combining fractions in such a procedure. However, since all fractions of interest are reassayed in detail alone, such a procedure rarely introduces errors.

B. Coupled Assay with Lactate Dehydrogenase

All enzymes used in this assay are available from Sigma, including the beef-heart cyclic nucleotide phosphodiesterase. Where appropriate, conditions for assay of the rat brain phosphodiesterase preparation described above are also given.

A second spectrophotometric technique that is also useful for assay of column effluents monitors absorbance at 340 nm rather than 265 nm. This method was developed in our laboratories, and is described next. It is not recommended for critical kinetic work, however, because of the large number of coupling enzymes used in this assay. This assay may prove useful for laboratories, familiar with spectrophotometric assays, that require a less elegant assay than that described in Section II for phosphodiesterase activity.

To 50 ml of 0.05 M imidazole buffer, pH 7.40, are added 372.8 mg of KCl, 23.3 mg of phosphoenolpyruvic acid (tricyclohexylammonium salt), and 0.2 ml of 2.5 M $MgCl_2$. This solution is stable for long periods of time if stored frozen. A second stock solution, the substrate mixture, is prepared by addition of 1.0 ml of a solution of 9.7 mg/ml ATP in 0.05 M imidazole buffer, pH 7.40, to 4 ml of this same buffer containing 68.3 mg of cyclic AMP. This mixture is also stable if stored frozen.

Each day, two simple working solutions are prepared. The first contains 5.0 ml of the $MgCl_2$-KCl-PEP buffer solution, to which has been added 5 µl of beef-heart lactate dehydrogenase (11 mg/ml) and 0.7 mg of DPNH. The second solution, the mixed enzyme dilution consists of 0.10 ml of the same imidazole buffer containing 20 µl of pyruvate kinase (from a stock Sigma preparation of 10 mg/ml) and 20 µl of myokinase (from a stock Sigma preparation of 5 mg/ml).

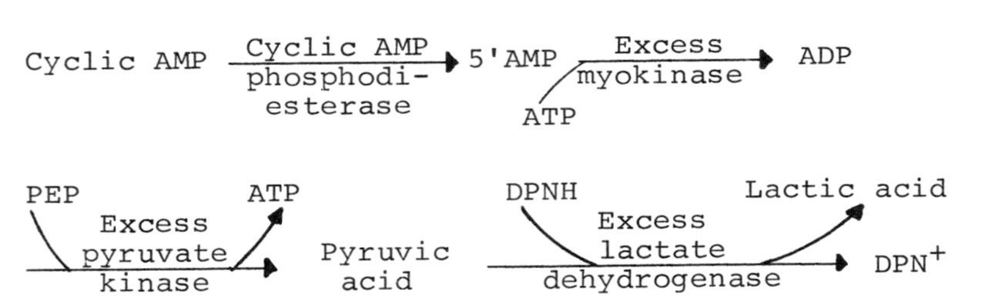

To determine phosphodiesterase activity, 0.4 ml of the first working solution is added to a semi-micro cuvette in a Gilford (or Beckman DU) spectrophotometer at a constant temperature of 25°C, if possible. To this solution are added 5 μl of the mixed enzyme dilution, 20 μl of the cyclic AMP-ATP mixture (final assay concentrations of 2 mM cyclic AMP and 0.4 μM ATP), and a suitable dilution of phosphodiesterase. Note that the final assay concentration of cyclic AMP may be varied considerably, if necessary. If the Sigma beef-heart cyclic nucleotide phosphodiesterase is used, the 5 mg of enzyme protein in the vial is dissolved in 2.0 ml of 0.05 M imidazole buffer, pH 7.40; 10 μl of this solution is suitable for assay. A suitable volume of the rat brain enzyme preparation described in Section II is 5 μl. The average initial rate of decrease of absorbance after mixing is proportional to phosphodiesterase activity.

To calculate specific activity for this assay, the absorbance change at 340 nm per min is divided by 15.55, which is the conversion factor based on the extinction coefficient of DPNH at 340 nm. This enzyme activity, in μmoles/min, is further divided by the amount of enzyme protein per assay, giving specific activity. For the Sigma beef-heart cyclic nucleotide phosphodiesterase, the specific activity cited was 0.39 μmoles/min/mg of protein and, using this assay, we found the specific activity to be 0.25 μmoles/min/mg of protein. Typical specific activity found for the rat-brain cyclic AMP phosphodiesterase preparation described in Section II was 0.10 μmoles/min/mg of protein.

It should be noted that the same buffers may be used in this assay as in the other spectrophotometric assay described in the preceding section.

IV. ACTIVITY STAIN FOR CYCLIC NUCLEOTIDE PHOSPHODIESTERASE IN POLYACRYLAMIDE GELS

We have found this activity stain, originally developed by Goren, et al. [25], particularly useful to locate phosphodiesterase activity after polyacrylamide gel electrophoresis. A solution is prepared containing 105.9 mg of $Pb(NO_3)_2$, 68.3 mg of $Mg(NO_3)_2$, and 1.03 g of tris base, in 90 ml of deionized, distilled water. The solution is adjusted to pH 7.0 with a concentrated solution of maleic acid. Note that either if $MgSO_4$ is used for the Mg^{2+} ion requirement, or if HCl is used for the titration, a precipitate of the appropriate insoluble salt occurs. After titration, the solution is diluted to 100 ml, and may be stored at room temperature. The reactions may be described as follows

$$\text{Cyclic AMP} \xrightarrow[\text{phosphodi-esterase}]{\text{Cyclic AMP}} \text{5'-AMP} \xrightarrow[\text{phosphatase}]{\text{Excess alkaline}} H_2PO_4^- \quad (+\ \text{adenosine})$$

$$\xrightarrow{Pb(NO_3)_2} Pb_3(PO_4)_2 \xrightarrow{(NH_4)_2S} \text{PbS (black-brown color)}$$

A suitable sample for electrophoresis of the Sigma beef-heart enzyme is 10 μl of a solution of 2.5 mg/ml in 0.05 M imidazole buffer, pH 7.4 (A 10 μl-aliquot of the rat brain phosphodiesterase described in Section I is also suitable.). Electrophoresis is performed exactly as described by Davis [26]. After electrophoresis of the phosphodiesterase preparation, we routinely slice the gels lengthwise in half, freehand, using a razor blade. Half of the gel may then be stained for protein by use of either Coomassie Blue [27] or Amido-schwartz [26] stain, and the other half is stained for phosphodiesterase activity. Alternatively, half may be stained for cyclic AMP phosphodiesterase activity and the other half for cyclic GMP phosphodiesterase activity. The half-gel to be stained for activity is placed in a 12 x 75 mm tube; the gel will slide in more easily if the tube is slightly wet. To this tube are added 3 ml of the tris-lead solution, 1 ml of the appropriate cyclic nucleotide (24 μmoles/ml) in deionized, distilled water, and finally 2.5 units of Sigma alkaline phosphatase in a volume of 5 μl or less. The tubes are then capped, mixed, and allowed to stand for 30 min at room temperature. After this incubation, the gels are removed and placed in similar tubes containing 5% ammonium sulfide, until the band of activity appears as a dark brown or black band in the gel. The gels are then washed and stored in water, in which the bands are stable for 3-5 days at room temperature, and slightly longer at 4°C. To obtain a permanent record of the results, the gels should either be photographed or scanned at 500 nm in a suitable scanning spectrophotometer.

V. OTHER METHODS OF ASSAY OF PHOSPHODIESTERASE ACTIVITY

Several other methods of assaying cyclic nucleotide phosphodiesterase activity have been reported, but will not be extensively discussed here. The most common of these is some modification of the original method of Butcher and Sutherland [1]; this method monitors inorganic phosphate released by the combined or sequential action of the cyclic nucleotide phosphodiesterase and snake venom 5'-nucleotidase. This procedure is applicable to any cyclic nucleotide, but it is restricted by the limitations of sensitivity of the assay for inorganic phosphate. Therefore, it becomes impractical for use as an assay for cyclic nucleotide phosphodiesterase activity at low substrate concentrations.

Another method that has found rather widespread use is based upon separation of the products of the reaction by paper or thin-layer chromatography [2,28, 29]. Two very useful solvents we have used are (1) 1.0 M ammonium acetate, pH 5.0-95% ethanol, 7:3, and (2) isobutyric acid-H_2O-tetramethylammonium hydroxide, 10% in water, 77:22:1. A more recent modification [30] of such chromatographic procedures uses high-pressure liquid chromatography to separate the reaction products. (See Chapter 3 for a more complete discussion of this technique). Use of radioactively-labeled cyclic nucleotides as substrates allows the use of chromatographic methods to assay phosphodiesterase activity over a wide range of substrate concentrations. In theory, these methods should be applicable to any cyclic nucleotide. The chief disadvantage to the method, however, is the relatively long and laborious procedure required for chromatography, as compared to the modified radiodisplacement assay, for instance. It should be noted that it is often important to trap the product of the cyclic nucleotide phosphodiesterase reaction with an exogenous unlabele product, to minimize further degradation of the product by contaminating enzymatic activities frequently found in the impure phosphodiesterase preparations generally used [29].

Trapping the product of the reaction is importan also when $Ba(OH)_2$-$ZnSO_4$ precipitation is used as a means of separating the products of the cyclic nucleotide phosphodiesterase reaction [31]. (This separation procedure is more fully discussed in Chapter 4 as applied to the adenylate cyclase reaction.) That is, unreacted cyclic AMP, but not 5'-AMP, the product, remains in solution after $Ba(OH)_2$-$ZnSO_4$ treatment. A disadvantage of this procedure is that it measures unreacted cyclic AMP present after incubation, rather than the product formed. It is, therefore, rather insensitive at low rates of reaction, which are the best conditions for detailed kinetic assays of enzymatic activities.

A titrimetric assay for cyclic nucleotide phosphodiesterase activity has also been developed [3 Since the cleavage of cyclic AMP releases a proton, then if the reaction is conducted in an automatic titration apparatus, the rate of addition of NaOH is proportional to enzymatic activity. A serious drawback of this procedure, however, is the very high (1 mM) substrate concentration required to perform the assay. This method cannot be used for assays of the activity of the low K_m phosphodiesterase.

The final types of assay to be discussed are cytochemical localization assays, applicable to tissue slices, cells, or chromatograms. Two of these [33,34] are based on the formation of insoluble lead sulfide from lead phosphate; the phosphate originates from cyclic AMP after cleavage by phosphodiesterase and a nucleotidase. A modification of this procedure has been described more extensively in Section IV. An alternative cytochemical procedure has been used by Monn and Christiansen [21] to detect multiple forms of cyclic AMP phosphodiesterase after starch-gel electrophoresis of various fractions from rat and rabbit tissue homogenate supernatants. After electrophoresis, the starch blocks are incubated in an enzyme mixture similar to that described in Section III.B; areas containing phosphodiesterase activity convert DPNH to DPN^+, which appears as a dark spot on a fluorescent background when viewed under ultraviolet light.

ACKNOWLEDGMENTS

We thank Mr. I. Rivkin, Mrs. M. B. Phillips, Mr. M. Rispoli, Mr. H. Goldenberg, and Miss S. G. Samaniego for their assistance in developing and assessing the assays described in this chapter. The kind cooperation and support of Dr. S. M. Hess is also gratefully acknowledged.

REFERENCES

[1] R. W. Butcher, and E. W. Sutherland, J. Biol. Chem., 237, 1244 (1962).

[2] O. M. Rosen, Arch. Biochem. Biophys., 137, 435 (1970).

[3] Y.-Y. Chang, Science, 161, 57 (1968).

[4] T. Okabayashi and M. Ide, Biochim. Biophys. Acta, 220, 116 (1970).

[5] J. G. Hardman and E. W. Sutherland, J. Biol. Chem., 240, 3704 (1965).

[6] U. Klotz and K. Stock, Naunyn-Schmiedebergs Arc Pharmakol., 269, 117 (1971).

[7] K. G. Nair, Biochemistry, 5, 150 (1966).

[8] W. Y. Cheung, J. Biol. Chem., 246, 2859 (1971).

[9] S. Chaykin, Biochemistry Laboratory Techniques, Wiley, New York, 1966, p. 17.

[10] A. G. Gornall, C. S. Bandawill, and M. M. David J. Biol. Chem., 177, 751 (1949).

[11] O. H. Lowry, N. J. Rosenbrough, A. L. Farr, and R. J. Randall, J. Biol. Chem., 193, 265 (1951).

[12] J. A. Beavo, J. G. Hardman, and E. W. Sutherland, J. Biol. Chem., 245, 5649 (1970).

[13] W. Y. Cheung, Biochemistry, 6, 1079 (1967).

[14] H. Sheppard and G. Wiggan, Mol. Pharmacol., 7, 111 (1971).

[15] M. Chasin, Fed. Proc., 30, 1268 (1971).

[16] W. R. Kukovetz and G. Poch, Naunyn-Schmidebergs Arch. Pharmakol., 267, 189 (1970).

[17] H. Sheppard and G. Wiggan, Biochem. Pharmacol., 20, 2128 (1971).

[18] G. Brooker, L. J. Thomas, and M. M. Appleman, Biochemistry, 7, 4177 (1968).

[19] W. J. Thompson and M. M. Appleman, Biochemistry, 10, 311 (1971).

[20] W. J. Thompson and M. M. Appleman, J. Biol. Chem., 246, 3145 (1971).

[21] E. Monn and R. O. Christiansen, Science, 173, 540 (1971).

[22] G. I. Drummond and S. Perrott-Yee, J. Biol. Chem., 236, 1126 (1961).

[23] Y. Lee, J. Biol. Chem., 227, 987 (1957).

[24] K. L. Smiley, A. J. Berry, and C. H. Suelter, J. Biol. Chem., 242, 2502 (1967).

[25] E. N. Goren, A. H. Hirsch, and O. M. Rosen, Anal. Biochem., 43, 156 (1971).

[26] B. J. Davis, Ann. N. Y. Acad. Sci., 121, 404 (1964).

[27] A. Chrambach, R. A. Reisfeld, M. Wyckoff, and J. Zaccari, Anal. Biochem., 20, 150 (1967).

[28] S. Jard and M. Bernard, Biochem. Biophys. Res. Commun., 41, 781 (1970).

[29] P. F. Gulyassy and R. L. Oken, Proc. Soc. Exptl. Biol. Med., 137, 361 (1971).

[30] S. N. Pennington, Anal. Chem., 43, 1701 (1971).

[31] G. Poch, Naunyn-Schmiedebergs Arch. Pharmakol., 268, 272 (1971).

[32] W. Y. Cheung, Anal. Biochem., 28, 182 (1969).

[33] T. R. Shanta, W. D. Woods, M. B. Waitzman, and G. H. Bourne, Histochemie, 7, 177 (1966).

[34] N. T. Florendo, R. J. Barrnett, and P. Greengard, Science, 173, 745 (1971).

Chapter 7

CYCLIC AMP-DEPENDENT AND CYCLIC GMP-DEPENDENT PROTEIN KINASES: PREPARATION AND ASSAY

Jyh-Fa Kuo

Department of Pharmacology
Yale University School of Medicine
New Haven, Connecticut 06510

I. INTRODUCTION

After the discovery of cyclic AMP-dependent protein kinase in mammalian skeletal muscle [1], this enzyme activity was found also in liver [2], and in no less than 50 other vertebrate [3] and invertebrate tissues [3,4]. In addition, the occurrence of cyclic GMP-dependent protein kinases, which are activated specifically by low concentrations of cyclic GMP rather than by cyclic AMP, was established in various lobster, insect, and mammalian tissues [4-6]. Cyclic AMP-dependent enzymes are invariably present in much higher concentrations than the cyclic GMP-dependent class of enzymes in mammalian tissues, whereas the relative level of the two classes of protein kinases in arthropod tissues is quite variable. Thus, for example, lobster-tail muscle and silkmouth larval body wall contain approximately equal levels of cyclic AMP-dependent and cyclic GMP-dependent protein kinases, lobster gill tissue contains almost exclusively cyclic AMP-dependent enzyme. The distribution pattern of the two classes of protein kinases clearly suggests an especially prominent role of the cyclic AMP system in mammals and of the cyclic GMP system in arthropoda; it also suggests the existence of distinctive functions for the two classes of protein kinases.

The reactions catalyzed by these enzymes are depicted as follows:

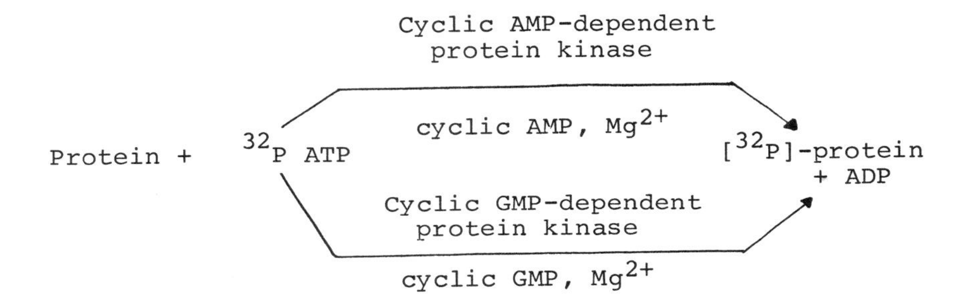

Since the natural substrates of cyclic AMP-dependent and cyclic GMP-dependent protein kinases in most tissues remain unknown, the commercial preparations of histone mixture, whose phosphorylation by ATP is effectively catalyzed by both classes of enzymes, is conveniently used as an artificial substrate to assay the activities of these enzymes from a wide variety of sources.

II. PREPARATION OF CYCLIC NUCLEOTIDE-DEPENDENT PROTEIN KINASES

A. From Tissues Containing Exclusively Cyclic AMP-Dependent Enzymes

All mammalian tissues and cells seem to belong to this category. Cyclic AMP-dependent protein kinases are prepared according to the following procedure, which is essentially the same as described elsewhere [1,3,7,8].

1. Preparation of Crude Extract

Fresh tissues, or frozen tissues (obtained from Pel-Freez) after they are thawed, are cut into small pieces (1-2 cm^3), and then homogenized with 3-4 vol of neutral 4-mM EDTA solution for 2 min in a Waring blendor. The homogenate is centrifuged at 27,000 x g for 30 min at 4°C. All procedures used for the purification of the enzymes are performed at 4°C. All buffers used in the succeeding steps of the purification contain 2 mM EDTA.

2. Acid Precipitation

The supernatant solution is adjusted to pH 4.8-5.1 by the dropwise addition of ice-cold 1-N acetic acid, with stirring. The pH that gives a maximal effect in bringing about the isoelectric precipitation of impure protein without removing the enzyme activity from the solution differs slightly among preparations of various tissues. A pH of 4.8, for example, is found to be the most effective in purifying enzymes from bovine brain and kidney, whereas it is too acidic for the preparations of bovine heart, testis and lung. In the latter tissue preparations, pH 5.0-5.1 is recommended to achieve an effective purification and a maximum recovery of the enzyme activity. After the enzyme is allowed to stand for 10 min, the precipitate is removed by centrifugation at 27,000 x g for 30 min. The pH of the clear supernatant solution is then adjusted to 6.8 with 1 M potassium phosphate buffer, pH 7.2.

3. Ammonium Sulfate Precipitation

Solid ammonium sulfate (33 g/100 ml) is added to the resultant solution from the previous step. After 20 min of stirring, the precipitate is collected by centrifugation at 27,000 x g for 20 min and dissolved in 6% of the crude extract volume of 5-mM potassium phosphate buffer, pH 7.0. The resulting solution is dialyzed overnight against 20 vol of the same buffer, with two changes of the buffer. After dialysis, the solution is centrifuged for 30 min, and the precipitate is discarded.

4. DEAE-Cellulose Chromatography

The enzyme solution is applied to a column (4 x 30 cm) of DEAE-cellulose (Sigma, medium mesh), which had been previously washed with 300-mM potassium phosphate buffer, pH 7.0, and equilibrated with 5-mM potassium phosphate huffer, pH 7.0. The amount of DEAE cellulose used is 56 g (dry weight) per g of protein. The flow rate is 2-4 ml per min. After the enzyme has been applied, the column is washed with 2 bed vol of 50-100-mM potassium phosphate buffer, pH 7.0, and the enzyme is then eluted from the column with 300-mM phosphate buffer, pH 7.0. (The concentration of the phosphate buffer used to selectively wash off the impure proteins without eluting the enzyme activity from the resin differs slightly among the preparations of different tissues. For example, 100-mM phosphate buffer satisfactorily washes off the protein impurities present in preparations of bovine brain and heart, but it is found to elute from the column both the protein kinase activity and the protein impurities in the preparations of bovine lung, stomach, and duodenum. For these tissues, 50-mM phosphate buffer is satisfactory for this step.) The active fractions eluted from the column with 300-mM phosphate are then pooled and dialyzed overnight against 20 vol of 5-mM phosphate, pH 7.0, with two changes of the buffer.

5. Calcium Phosphate Gel Treatment

Calcium phosphate gel is prepared according to

the method of Keilin and Hartree [9]. (To a 20-liter glass jar containing 2 liters of deionized water and 500 ml of 0.6 M $CaCl_2$ is slowly added, with stirring, 500 ml of 0.4 M Na_3PO_4, followed by 1 N acetic acid until the pH is 7.3. The jar is then filled with water. The precipitate is washed six times with distilled water, and is finally suspended in an appropriate volume of water to give a thick gel suspension of 30 mg dry weight per ml of gel.) Calcium phosphate gel is added dropwise to the enzyme solution obtained from the previous step, with stirring. The amount of the gel used is 2-2.5 g (dry weight) per g of protein. The precipitate is collected by centrifugation at 3300 x g for 5 min. The precipitate is suspended in 30-mM potassium phosphate buffer, pH 7.0 (the volume of the buffer used is 20% of the original volume of the dialyzed enzyme solution), and then recentrifuged. This washing procedure is done three times. The protein kinase activity is finally eluted from the gel with 300-mM phosphate buffer, pH 7.0, the volume of the buffer used being 10% of the dialyzed enzyme solution from the ammonium sulfate step. The elution step is done three times. The supernatant solutions are combined and dialyzed overnight against 20 vol of

5-mM phosphate buffer, pH 7.0, with two changes of the buffer.

Owing to the presence of inhibitory substance(s), the protein kinase activity and the stimulation of the enzyme by cyclic AMP seen in the crude tissue extract are rather low. The activity of the enzyme and its stimulation by cyclic AMP increase, however, as the enzymes are purified in the succeeding steps. In a great majority of the tissue preparations, the measurable enzyme activity is found to be highest in the ammonium sulfate step, accompanied by the greatest or near greatest stimulation by cyclic AMP. The preparations from the ammonium sulfate, DEAE cellulose, and calcium phosphate steps are satisfactory for use in studies concerning the occurrence of the enzymes in tissues or the catalytic properties of the enzymes.

The enzymes, prepared from various tissues according to the procedure described here, are stimulated maximally (4- to 30-fold) by 0.2-0.5 μM cyclic AMP, or by 25-100 μM cyclic GMP (see Section III for the assay conditions). The recovery of the enzyme activity from various tissues is estimated to be from as low as 5% from some tissues to as high as 95% from others.

B. From Tissues Containing Exclusively Cyclic GMP-Dependent Enzymes

The fat body of cecropia silkmoth pupae (or larvae) is the only tissue found to date to contain exclusively cyclic GMP-dependent protein kinase [4]. The procedure for the partial purification of the enzyme [4] from the insect fat body should also be applicable to other tissues, perhaps with minor modifications.

The first three steps (i.e., extraction, pH precipitation, and ammonium sulfate fractionation) are essentially the same as described in subsection A, preceding for tissues containing exclusively cyclic AMP-dependent protein kinases. A pH of 4.8, however, is the most satisfactory for the insect fat body preparation. The enzyme preparation thus obtained is very active, and is stimulated about 4- to 7-fold by 0.5-μM cyclic GMP or by 25-μM cyclic AMP.

The enzyme can be readily purified further (about 2-fold) by a simple step, as described below. To the dialyzed enzyme solution obtained from the ammonium sulfate step (II.A.3) is added calcium phosphate gel. The amount of the gel added is about 0.6 mg of dry gel per 1.0 absorbance unit (at 280 nm) of the enzyme solution. After the mixture is gently stirred for 5 min in ice, it is centrifuged at 27,000 x g for 10 min, and the precipitate is discarded. More than 95% of the cyclic GMP-dependent protein kinase activity present in the enzyme solution from the previous step is recovered in the supernatant solution. The sensitivity to, and the magnitude of activation by cyclic GMP, seen in the resultant enzyme preparation is comparable to the preparation from the previous step.

C. From Tissues Containing Both Classes of Enzymes

Most of the arthropod tissues belong to this category. The two classes of protein kinases are purified and separated from each other as follows [4,5].

1. Method 1

About 300 g of tail muscle from three live Maine lobsters (obtained from a local fish market) is used as the starting material. The first three steps (i.e., extraction, acid precipitation, and ammonium sulfate fractionation) are the same as for the preparation of cyclic AMP-dependent enzyme described in Section A. A pH of 4.9 is optimum for the acid precipitation step for the preparation of protein kinases from lobster-tail muscle. The dialyzed enzyme solution obtained from the ammonium sulfate step (containing both cyclic AMP-dependent and cyclic GMP-dependent enzymes) is then applied to a column (2.4 x 14 cm) of DEAE cellulose previously washed with 300-mM phosphate buffer, pH 7.0, and equilibrated with 5-mM phosphate buffer, pH 7.0. Protein is eluted from the column by a stepwise application of 100 ml each of 5-, 50-, 100-, and 200-mM potassium phosphate buffer, pH 7.0. Cyclic GMP-dependent protein kinase activity is associated with the protein peak eluted by 5-mM phosphate buffer, and cyclic AMP-dependent protein kinase activity is found in the protein peak eluted by the 50-mM phosphate buffer. Although no significant cyclic nucleotide-dependent protein kinase activity is found in the protein peaks eluted by 100- and 200-mM phosphate buffer when lobster muscle is used, these buffers may be useful for eluting kinases from other tissues. The active fractions eluted with 5-mM and 50-mM phosphate are separately pooled; they are designated as peak 1 and peak 2, respectively. The possible contamination of cyclic AMP-dependent enzyme activity in peak 1 and cyclic GMP-dependent activity in peak 2 are minimal. The preparations of the two classes of protein kinases at this stage of purity are satisfactory for studies of their catalytic properties. Generally, under the assay conditions to be described later, the protein kinase activity in peak 1 is activated 4-fold by 0.5-μM cyclic GMP, whereas it is activated only 0.5-fold by 0.5-μM cyclic AMP. The protein kinase activity in peak 2 is activated, on the other hand, 5-fold by 0.5-μM cyclic AMP and about 1.8-fold by 0.5-μM cyclic GMP.

Cyclic GMP-dependent protein kinase in peak 1 can be readily purified further (2-fold) by a calcium phosphate gel treatment, as described in Section B for the same enzyme activity prepared from insect fat body.

2. Method 2

The first three steps of this method are the same as in Method 1. The starting materials, usually 1-10 g of fresh arthropod (e.g., lobster and insect) tissue, are homogenized with a small glass homogenizer, or an Omni-Mixer (with or without the attachment for small samples). In this method, the calcium phosphate gel step replaces the DEAE-cellulose chromatography step in Method 1 for the separation of cyclic AMP-dependent and cyclic GMP-dependent protein kinases. This calcium phosphate gel step has certain advantages over the DEAE-cellulose step; it can be used, with a minimum loss of the enzyme activity, for small-scale enzyme preparations from tissues of limited availability in a large quantity. Furthermore, the calcium phosphate gel step can be used as effectively as the DEAE-cellulose step for a large-scale enzyme preparation from some tissues, such as lobster-tail muscle.

To the dialyzed enzyme solution from the ammonium sulfate step is added calcium phosphate gel [30 mg of dry weight per ml of suspension; about 5 mg of dry gel per 1.0 absorbance unit (at 280 nm of the enzyme)]. After gentle stirring of the mixture for 5 min in ice, the gel is recovered by centrifugation. The supernatant fluid contains little or no cyclic nucleotide-dependent protein kinase activity; it is discarded. Cyclic GMP-dependent protein kinase activity is eluted first from the gel by suspending it in an appropriate volume of 50-mM phosphate buffer containing 2 mM EDTA, pH 7.0, and stirring gently for 5 min in ice. The gel, collected by centrifugation, is eluted once again with the same buffer. The two eluates obtained with 50-mM phosphate buffer are pooled and designated as gel eluate 1. For recovery of cyclic AMP-dependent protein kinase activity, appropriate volumes of 200-mM phosphate containing 2 mM EDTA, pH 7.0, are then added to the gel, and the elution procedure is performed twice as described above. The two eluates obtained with 200-mM phosphate are pooled, and are designated as gel eluate 2. The volumes of 50-mM phosphate and 200-mM phosphate used are such that the absorbancies of the eluates at 280 nm are between 0.2 and 1.5, by eluting with between 20 and 30% of the volume of dialyzed enzyme preparation. This procedure gives effective and simple separation of the two types of protein kinase activity. The entire procedure of calcium phosphate gel treatment is performed even for those tissue preparations whose ammonium sulfate fractions (and earlier fractions) indicate only the possible occurrence of either cyclic AMP-dependent or cyclic GMP-dependent protein kinase activity.

III. ASSAY FOR PROTEIN KINASE ACTIVITY

A. Preparation of Reagents

1. Histone Mixture

Dissolve 100 mg of histone mixture (calf thymus, Schwarz/Mann) in 20 ml of 0.01 N HCl. Neutralize the solution with 5 N NaOH, and add water to make a final volume of 50 ml. The resultant histone solution should be clear. If not, check to be sure the pH is between 6.5 and 7.0.

2. [γ-^{32}P]-ATP

In our laboratory, this is prepared from ortho-^{32}P according to the method of Post and Sen [10]. It can be obtained, however, from commercial sources (e.g., ICN, New England Nuclear, etc.).

3. Trichloroacetic Acid-Sodium Tungstate-H_2SO_4 Precipitating Solution

Slowly dissolve 60 g of NaOH pellets, with stirring, in 5 liters of 5% trichloroacetic acid (the pH of the resultant solution should be about 2); then dissove 12.5 g of $NaWO_4 \cdot 2H_2O$ in this solution. Adjust the pH to 2 with 100% trichloroacetic acid, and finally add 10 ml of concentrated (36 N) H_2SO_4. The resultant solution remains clear for at least several months.

B. Standard Assay Procedure

For the assay of cyclic AMP-dependent protein kinase activity, the following components are added to the incubation tubes (13 x 100 mm) in a final volume of 0.2 ml, in the order given.

Solution	Volume (ml)
Sodium acetate buffer, 0.1 M, pH 6.0	0.100
Magnesium acetate (2 μmoles)	0.020
Cyclic AMP (0.01-1 nmole), or water	0.020
Histone mixture (40 μg)	0.020
Other additive, if any, or water	0.014
Cyclic AMP-dependent protein kinase (20-200 units)	0.020
[γ-^{32}P] ATP (1 nmole)	0.006

For the assay of cyclic GMP-dependent protein kinase, this enzyme preparation (20-200 units) and cyclic GMP (0.01-1 nmole) replace cyclic AMP-dependent enzyme and cyclic AMP, respectively, in the incubation system given above. The tubes are kept in ice during the addition of the reagents. One unit of enzyme is defined as the amount of enzyme that transfers 1 pmole (10^{-12} mole) of ^{32}P from [γ-^{32}P] ATP to histone (or other protein substrates) in 5 min under the assay conditions. The reaction is begun by the addition of radioactive ATP (containing about 1.5-2.2 x 10^6 cpm), which is conveniently delivered from a 100-μl Hamilton microsyringe (with a fused needle) attached to a repeating dispenser. The tubes are incubated for

5 min at 30°C, with shaking, and the reaction is terminated by the addition of 2 ml of ice-cold trichloroacetic acid-tungstate-H_2SO_4 (hereafter referred to as the precipitating solution). Bovine serum albumin (0.6%, 0.2 ml) is added as a carrier protein, and the contents of the tubes are mixed by vigorous addition of another 2 ml of the precipitating solution. The mixture is centrifuged in a PR-6 International Refrigerated Centrifuge (equipped with a rotor of 48-tube capacity) at about 2000 rpm, and the supernatant solution is removed by aspiration. The precipitate is dissoved in 0.1 ml of 1 N NaOH, and 2 ml of the precipitating solution are added. The procedure of centrifuging, removing the supernatant, dissolving the precipitate in alkali, and reprecipitating the protein is repeated once more. The protein is finally collected by centrifugation and dissolved in 0.1 ml of 1 N NaOH, and the radioactivity is counted in 6 ml of scintillation fluid (made by dissolving 8 g of Omnifluor, purchased from New England Nuclear, in 1 liter of toluene and 1 liter of ethylene glycol monoethyl ether). Under the assay conditions, the activity of the enzymes is linear for at least 8 min of incubation.

A typical example of the activity of the two classes of protein kinases obtained under various incubation conditions is given in Table I. Histones used were obtained from Schwarz/Mann, and protamine and casein were from Sigma. Results qualitatively similar to those in Table I have been obtained for the individual classes of protein kinases prepared from diverse sources.

IV. CONCLUDING REMARKS

Cyclic AMP-dependent and cyclic GMP-dependent protein kinases can be purified individually and separated from each other in extracts of various tissues by the general procedure described in the preceding sections. The enzyme preparations thus obtained are satisfactory for studies concerning the distribution of the enzymes in tissues and cells, or the catalytic properties of the enzymes. In studies for which highly purified preparations are required, the protein kinases can be further purified according to the procedure described for bovine brain [7] and rabbit skeletal muscle [11], perhaps with only minor modifications.

The assay system for measurement of the activity of protein kinase has been modified from the one

TABLE 1

Activity of Cyclic Nucleotide-Dependent Protein Kinase Assayed under Various Incubation Conditions

Components in the incubation system	Relative Activity of	
	Cyclic AMP Dependent Enzyme (bovine heart)	Cyclic GMP Dependent Enzyme (lobster muscle)
Complete	100	100
Minus individual cyclic nucleotides, 0.5 μM	7	23
Minus histone mixture, 40 μg	3	15
Minus Mg^{2+}, 10 mM	5	7
Complete, substitute individual cyclic nucleotide, 0.5 μM		
With cyclic AMP, 0.5 μM	100	32
With cyclic AMP, 5 μM	103	65
With cyclic GMP, 0.5 μM	12	100
With cyclic GMP, 5 μM	25	107
Complete, substitute histone mixture, 40 μg, with arginine-rich histone, 40 μg	125	118
With lysine-rich histone, 40 μg	32	80
With protamine, 40 μg	25	50
With casein, 100 μg	8	18
Complete, substitute Mg^{2+}, 10 mM,		
With Co^{2+}, 10 mM	95	230
With Mn^{2+}, 10 mM	48	180
With Ca^{2+}, 10 mM	5	8

described earlier by Walsh, et al. [1]. We found that protein kinases (both cyclic AMP-dependent and cyclic GMP-dependent) are much more active in acetate buffer than in glycerol-phosphate buffer, that histone is a much better substrate (about 30 to 100 times more reactive) than casein, that theophylline and NaF are unnecessary under our assay conditions and, therefore, are omitted, and that a low concentration of ATP usually yields a higher stimulation of the enzyme activity by the added cyclic nucleotides. Trichloroacetic acid-tungstate-H_2SO_4 solution is used for quantitative recovery of histone from the reaction mixture, since histone is partially soluble in 5% trichloroacetic acid.

The activity of cyclic nucleotide-dependent protein kinases provides a sensitive means for assaying cyclic AMP and cyclic GMP. The protein kinase catalytic methods for assaying cyclic nucleotides are based upon the ability of low concentrations of cyclic AMP [12-14] and cyclic GMP [14,15] to activate cyclic AMP-dependent and cyclic GMP-dependent protein kinase, respectively. We found that the limit of sensitivity of the procedure for the assay of cyclic AMP with bovine-heart cyclic AMP-dependent protein kinase is about 0.2 pmole, and that for the assay of cyclic GMP with lobster cyclic GMP-dependent protein kinase is about 0.5 pmole.

ACKNOWLEDGMENT

This work, based on several original research articles published jointly with Drs. Paul Greengard and G. R. Wyatt, was supported in part by Grant HE-13305 from the United States Public Health Service and by Grant G-70-31 from the Life Insurance Medical Research Fund. The author is recipient of a Research Career Development Award (1 K4 GM-50165) from the United States Public Health Service.

REFERENCES

[1] D. A. Walsh, J. P. Perkins, and E. G. Krebs, J. Biol. Chem., 243, 3763 (1968).

[2] T. A. Langan, Science, 162, 579 (1968).

[3] J. F. Kuo and P. Greengard, Proc. Natl. Acad. Sci. U. S., 64, 1349 (1969).

[4] J. F. Kuo, G. R. Wyatt, and P. Greengard, J. Biol. Chem., 246, 7159 (1971).

[5] J. F. Kuo and P. Greengard, J. Biol. Chem., 245, 2493 (1970).

[6] P. Greengard and J. F. Kuo, in Role of Cyclic AMP in Cell Function (P. Greengard and E. Costa, eds.), Raven Press, New York, 1970, pp. 287-306.

[7] E. Miyamoto, J. F. Kuo, and P. Greengard, J. Biol. Chem., 244, 6395 (1969).

[8] J. F. Kuo, B. K. Krueger, J. B. Sanes, and P. Greengard, Biochim. Biophys. Acta, 212, 79 (1970).

[9] D. Keilin and E. F. Hartree, Biochem. J., 49, 88 (1951).

[10] R. L. Post and A. K. Sen, in Methods in Enzymology, Vol. X (R. W. Eastabrook and M. E. Pullman, eds.), Academic Press, New York, 1967, pp. 773-774.

[11] E. M. Reimann, D. A. Walsh, and E. G. Krebs, J. Biol. Chem., 246, 1986 (1971).

[12] J. F. Kuo and P. Greengard, J. Biol. Chem., 245, 4067 (1970).

[13] W. B. Wastila, J. T. Stull, S. E. Mayer, and D. A. Walsh, J. Biol. Chem., 246, 1996 (1971).

[14] J. F. Kuo and P. Greengard, in Advances in Cyclic Nucleotide Research Vol. 1 (P. Greengard, G. A. Robison, and R. Paoletti, eds), Raven Press, New York, in press.

[15] J. F. Kuo, T. P. Lee, P. L. Reyes, K. G. Walton, T. E. Donnelly, and P. Greengard, J. Biol. Chem., 247, 16 (1972).

Part 3

THE USE OF WHOLE CELL SYSTEMS

Chapter 8

LIPOCYTE AND ADRENAL CELL SUSPENSIONS

Charles A. Free

Department of Biochemical Pharmacology
The Squibb Institute for Medical Research
New Brunswick, New Jersey 08903

I. INTRODUCTION

Suspensions of isolated cells provide a valuable experimental tool for the study of cyclic AMP-mediated metabolic processes, and are of particular utility in the search for pharmacologically active agents that affect these processes. These preparations of isolated, and yet intact, cells allow the examination of agents in integrated metabolic systems, in distinction to the more limited events or reactions that are generally studied in cell-free homogenates or partially-purified enzyme fractions. In comparison with preparations such as sliced, perfused, or superfused tissues, on the other hand, isolated cells offer an advantage of uniformity in the preparation of replicate samples. Moreover, cell suspensions compare favorably with cells *in vivo* with respect to accessibility to effector agents, in contrast to the conditions prevailing in excised tissues that no longer contain functional vascular apparatus.

This chapter will deal with two widely used kinds of isolated cells, prepared from the epididymal fat pad and the adrenal gland of the rat. For these cell types, the isolated fat cell (or lipocyte) and the isolated adrenal cell, substantial evidence has been gathered to implicate cyclic AMP as a mediator of their respective, hormonally activated lipolytic or steroidogenic activities [1]. In addition to presenting methods for the preparation and experimental use of the isolated cells, this chapter will review their applications in the study of effector agents. In the latter instances, the emphasis will be placed on those agents that appear to act directly at sites of synthesis, action, or degradation of cyclic AMP.

II. LIPOCYTES

Rodbell [2] made the original contribution to the development of isolated-cell systems with his observation that treatment of epididymal fat tissue with bacterial collagenase resulted in a suspension of individual cells. Upon centrifugation, a homogenous fraction of buoyant fat cells was readily separated from the sedimenting stromal-vascular cells. These isolated fat cells retained most of the metabolic properties and responsiveness to hormones [2] of the parent adipose tissue. The excellence of this technique for the preparation of lipocytes has been demonstrated by its use in many laboratories, with relatively little modification, to the present. The following procedures for preparation and use of lipocytes are based upon those of Rodbell, with several modifications that have been introduced by our laboratory in the course of more than four years of experimentation with these cells.

A. Preparation and Incubation

A Krebs-Ringer phosphate-albumin buffer, pH 7.4, is used for most of the steps in both preparation and incubation of isolated fat cells. This buffer is prepared according to DeLuca and Cohen [3], but contai half the specified concentration of $CaCl_2$, no glucose, and either bovine albumin (Fraction V, Miles Laborator Inc.; 3 g/100 ml) or human albumin (Fraction V, E. R. Squibb and Sons; 2 g/100 ml). The albumin functions as a stabilizing agent for the cells, and also as a trapping agent for fatty acids released during lipolys To prepare the buffer, combine: 300 ml NaCl, 0.154 M; 12 ml KCl, 0.154 M; 9 ml $CaCl_2$, 0.055 M; 60 ml NaH_2PO_4 0.114 M (adjusted to pH 7.4 with HCl); 3 ml $MgSO_4$, 0.154 M; and 11.5 g bovine albumin (see preceding). The pH of the combined solutions should be checked and if necessary, readjusted to pH 7.4 with HCl.

The collagenase used for the digestion of adipose tissue is a crude preparation (Code CLS, Worthington Biochemical Corp.) derived from *Clostridium histolytic*

The properties of this crude enzyme appear to derive in part from other proteases present as impurities, and some lots have proven more satisfactory than others with respect to the activity and properties of the resulting lipocyte preparations. It is recommended that 3 to 6 lots be compared in order to select the one with the most satisfactory activity (see Section II.C).

At all stages of preparation and incubation, lipocytes tend to lyse on contact with glass surfaces. Accordingly, plastic or siliconized glass utensils are used throughout. For siliconization of glassware, treatment with Clay-Adams "Siliclad" is recommended: expose the clean, dry glass surface to 1% Siliclad solution in H_2O, rinse in H_2O, and dry at room temperature.

Male Sprague-Dawley rats of 100-250 g, fed and watered *ad libitum*, are the source of epididymal adipose tissue. The animals are killed by a blow on the head or by decapitation. Distal portions of the epididymal fat pads are removed, transferred to a tared container of 0.154 M NaCl, and weighed (a yield of 0.5-1.5 g tissue/rat may be expected, depending upon the size of the animals). The rinsed tissue is removed from the saline, blotted on a paper towel, minced with scissors, then placed into a 25-ml plastic scintillation-counting vial. To the minced tissue in the vial are added Krebs-Ringer phosphate-albumin buffer (1 ml/g tissue) and collagenase (5 mg/g tissue). The vial is capped, placed on its side, and agitated in a New Brunswick Gyrotory Shaker at 160 cycles/min for 1 hr at 37°C. After the collagenase treatment, the resulting cell suspension is decanted into a 40-ml conical (glass, graduated) centrifuge tube and centrifuged for 1 min at 400 x g. Upon inclining the tube, the infranatant buffer and pellet of stromal-vascular cells may be readily removed and discarded by aspiration. The lipocytes are washed two more times by resuspension in the original volume of buffer and recentrifugation at 400 x g. After the final wash, the cells are resuspended by the addition of 6 vol of buffer and minimal agitation for about 2 sec on a Vortex mixer. The expected yield of lipocyte suspension is 7.5-10 ml/g adipose tissue.

As many as 60 individual incubation mixtures are included in a routine lipolysis assay. The incubation mixtures are prepared by combining, in 25-ml plastic scintillation-counting vials, 1.0 ml of lipocyte suspension (a 5-ml siliconized wide-mouth pipet is suitable for dispensing aliquots of the suspension), test substances in volumes up to 0.2 ml, and Krebs-Ringer phosphate-albumin buffer, to a final volume of 2.5 ml. The open vials are incubated at 37°C for 1 hr in a New Brunswick Gyrotory Shaker oscillating at 160 cycles/min, and covered to minimize evaporation from the vials. After the incubation period, the reactions are terminated by the addition of 0.5 ml of 3 M $HClO_4$ to each vial, followed by gentle agitation for 30 sec. The mixtures, still in the vials, are then centrifuged for 3 min at 1000 x g. After the centrifugation, the aqueous supernatant fractions are decanted into clean sample vials; any fat droplets remaining on the aqueous extracts after the decantation step should be removed by aspiration. Aliquots of these clear aqueous extracts are used for the subsequent determination of glycerol released during incubation of the lipocytes.

B. Determination of Glycerol

The determination of glycerol released after the intracellular conversion of fat into free fatty acids and glycerol provides a highly sensitive and specific method for measuring the rate of lipolysis in isolated fat cells. Glycerol concentrations in the $HClO_4$-treated incubation mixtures are determined by a fluorometric method that measures NADH formed during the conversion of glycerol into dihydroxyacetone phosphate in the presence of glycerol kinase, ATP, and glycerol-3-phosphate dehydrogenase. A hydrazine buffer is used in the reaction mixture to trap dihydroxyacetone phosphate, assuring quantitative oxidation of the phosphorylated glycerol. Both manual and automated versions of the assay may be performed with the following reagents [4]:

(1) hydrazine·HCl buffer, 1 M, pH 9.6;

(2) enzyme reagent, containing

(a) 15 ml of sodium acetate (0.133 M), disodium ethylene-diamine tetraacetic acid (0.0133 M), $MgSO_4$ (0.1 M);

(b) 3 ml of ATP, 50 mg/ml in 1% $NaHCO_3$;

(c) 2 ml of NAD^+, 50 mg/ml;

(d) 0.2 ml of glycerol phosphate dehydrogenase suspension, 10 mg/ml (EGAQ, Boehringer);

(e) 0.02 ml of glycerol kinase suspension, 5 mg/ml (Boehringer).

The use of Boehringer preparations of glycerol phosphate dehydrogenase and glycerol kinase is advised, inasmuch as these preparations, unlike those from other suppliers, do not contain glycerol added as a stabiliz agents.

Manual determinations of glycerol are performed by mixing 1.8 ml of hydrazine·HCl buffer with 0.2 ml of sample, and then adding 1.0 ml of enzyme reagent. After 10 min (an interval sufficient for completion of the reactions at room temperature) fluorescence intensities of the solutions are measured at excitation and emission wavelengths of 350 and 460 nm, respectively.

The manual determination of glycerol may also be performed spectrophotometrically, employing absorbance at 340 nm, rather than fluorescence, to measure NADH formation. In other respects the fluorometric and spectrophotometric procedures are identical. The spectrophotometric method has the disadvantage of reduced sensitivity; it allows the measurement of minimum quantities of about 50 nmoles of glycerol in a $HClO_4$-treated incubation mixture. The fluorometric procedure, using a spectrofluorometer of sensitivity comparable to that of the Farrand Mark I, measures glycerol quantities as low as 5 nmoles per incubation mixture.

In our laboratory, glycerol determinations are performed by the automated fluorometric method of Ko and Royer [4], with the following flow rates adapted for use with an Aminco-Bowman spectrophotofluorometer equipped with an Aminco-Bowman 1-ml flow cell: glycerol sample, 0.10 ml/min; hydrazine·HCl buffer, 2.9 ml/min; air, 1.20 ml/min; enzyme reagent, 0.42 ml/min; and flow-cell output, 2.9 ml/min. Fifty samples per hour are analyzed.

In the manual and automated procedures alike, glycerol quantities are calculated from standard curves. These curves are obtained by inclusion of a blank and glycerol standards of 0.01-0.50 μM, all in 0.5 M $HClO_4$, as an integral part of every set of determinations. In adddition, it is necessary that agents tested in lipocyte incubation mixtures be tested in parallel as components in a glycerol standard the latter test detects possible interference of agents through indirect effects, such as inhibition of glycerol phosphate dehydrogenase or glycerol kinase, or through more direct effects such as self-fluorescence or quenching.

C. Properties of Lipocytes

In lipocytes that are prepared and incubated under the conditions described in the preceding, the basal rate of glycerol formation is generally less than 50 nmoles/hr. In contrast, the maximum lipolytic rate resulting from stimulation by various agents ranges from 750 to 1500 nmoles/hr. Among the agents that activate lipolysis in the isolated fat cells are several hormones, including ACTH, thyroid-stimulating hormone, and epinephrine [2]. These hormones, as well as glucagon and luteinizing hormone, stimulate the activity of lipocyte adenylate cyclase [5] and raise cyclic AMP concentrations in lipocytes [6], supporting the hypothesis that cyclic AMP mediates the lipolytic actions of these hormones. Additional support for this hypothesis is provided by the lipolytic activities of cyclic AMP and its N^6,2'-O-dibutyryl analog, both of which are capable of maximally stimulating lipolyis. Figure 1 illustrates the relative activations of lipolysis by ACTH, epinephrine, and the two cyclic nucleotides. Each of the four agents activates the cells to the same maximum lipolytic response, although the activating potencies vary over a wide range. The respective concentrations for half-maximal activation are: ACTH, 0.013 μM; epinephrine, 0.22 μM; dibutyryl cyclic AMP, 500 μM; and cyclic AMP, 8500 μM.

III. ISOLATED ADRENAL CELLS

The initial technique for preparation of isolated adrenal cells was developed by Kloppenborg et al. [8], and was based upon collagenase digestion

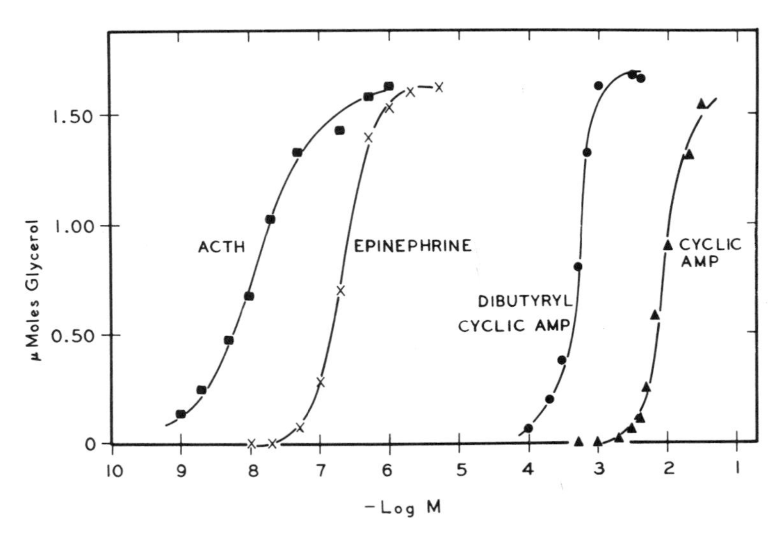

Fig. 1. Glycerol production by lipocytes in response to ACTH, epinephrine, dibutyryl cyclic AMP, and cyclic AMP. All points represent individual incubations within a single experiment. (Reprinted from Ref. [7], p. 3787, by permission of the American Chemical Society.)

and gentle mechanical disruption to free the cells from adrenal connective tissue. There have been many modifications of this technique, all of which use adrenal glands from rats. Although one early variatio [9] used a mixture of five enzymes, the remaining methods may be divided into two general groups, based upon the use of collagenase [8,10-12] or trypsin [13-16] for the digestion of connective tissue. Cells prepared by all these methods are highly sensitive to hormonal activation, demonstrating maximum secretion of corticosterone at ACTH concentrations of about 1 International milliunit/ml or lower (1 mU/ml = 2 nM).

The following method for preparing isolated adrenal cells uses a collagenase treatment of rat adrenal tissue, and yields cells with properties similar to those of cells prepared by the related techniques [8,10-12]. This isolated-cell preparation demonstrates good levels of maximum steroidogenic activity and sensitivity to ACTH. Moreover, the cells are capable of appropriate responses, via inhibition or stimulation of corticosterone secretion, to agents that act on sites of synthesis, degradation, or action of cyclic AMP.

A. Preparation and Incubation

The Krebs-Ringer bicarbonate-albumin-glucose buffer used for the preparation and incubation of isolated adrenal cells is prepared according to DeLuca and Cohen [3], except that it contains 1.5 times the specified concentration of $CaCl_2$, with additions of bovine albumin (Fraction V, Miles

Laboratories, Inc.; 3 g/100 ml) and glucose (0.2 g/100 ml). To prepare the buffer, combine: 300 ml NaCl, 0.154 M; 12 ml KCl, 0.154 M; 9 ml $CaCl_2$, 0.165 M; 3 ml KH_2PO_4, 0.154 M; 3 ml $MgSO_4$, 0.154 M; 63 ml $NaHCO_3$, 0.154 M (gassed with CO_2 for 1 hr); 0.78 g glucose; and 11.7 g albumin.

Crude collagenase (Code CLS) is obtained from Worthington Biochemical Corp. As in the case of collagenase lots used for the preparation of lipocytes, only about half are satisfactory; it is suggested that 3 to 6 lots be compared to assure selection of one that produces adrenal cells with optimum steroidogenic properties (see Section III. C).

Certain experimental conditions for the incubation of adrenal cells require the addition of very small quantities of ACTH. In order to facilitate the daily preparation of dilute solutions of the hormone, as well as to avoid waste, the preparation of a low-potency ACTH-albumin mixture is recommended. Such a preparation may be obtained by lyophilization of a solution of ACTH (porcine ACTH, Sigma Chemical Co.; about 100 International Units/mg) and bovine albumin. Lyophilization of a 50-ml solution of ACTH (30 μg/ml) and bovine albumin (40 mg/ml), for example, yielded 2.0 g of a stable powder of about 60 mU/mg. The

ACTH preparations described in this chapter are standardized against the Third International Working Standard for Corticotrophin (obtained from the World Health Organization Laboratory for Biological Standards, National Institute for Medical Research, London, England) and their potencies, like that of the standard, are expressed in terms of the International Unit [17]. The International Unit is based originally on a subcutaneous bioassay, *in vivo*, and should not be confused with "i.v." units (of about three times the activity) that are based on the use of an intravenous route of bioassay.

Male Sprague-Dawley rats of 100-250 g, fed and watered *ad libitum*, are used as the source of adrenal tissue. The following procedure describes the volumes and equipment for a typical cell preparation, using the adrenals of 16 rats (sufficient for about 30 cell incubations). After the rats are killed by decapitation, the adrenals are removed and collected in cold buffer. Decapsulation is accomplished by transfer of the glands to a sheet of cellophane, bisection with a scalpel, and then, by gentle external pressure, extrusion of the decapsulated portions of the glands from the surrounding capsules. The lighter-colored capsular tissue consists mainly of glomerulosar cells of relatively low corticosteroidogenic activity, and comprises roughly 20% of the mass of the gland. The decapsulated adrenal tissue is transferred to a 25-ml Erlenmeyer flask containing buffer (5 ml per 32 adrenals) and collagenase (5 mg/ml). Digestion is conducted for 1 hr at 37°C, under 95% O_2-5% CO_2, in a New Brunswick Gyrotory Shaker oscillating at 120 cycles/min.

After the digestion period, the original fragmen remain visible in the tissue suspension; these pieces are disrupted by gentle and repeated drawing of the mixture into a Pasteur pipet (about 20 to 25 2-ml excursions generally provide an optimum degree of cellular dispersion). The suspended cells are collec by centrifugation at 4°C for 10 min at 480 x g, followed by two cycles of washing and recentrifugatic in 10 ml of cold buffer. The washed cell pellet is then resuspended in 32 ml of buffer (1 ml per adrenal and filtered through a stainless-steel sieve with perforations of 0.2 mm to remove any large particles of undigested tissue (the support screen in a Millipore Swinny hypodermic adapter unit is suitable for this filtration).

Adrenal cells are incubated in 25-ml Erlenmeyer flasks in volumes of 2.5 ml. To each flask are addec in sequence, buffer, solutions of test substances in volumes of up to 0.1 ml, and 1 ml of adrenal cell suspension, the latter added gently from a 5-ml, wide-mouth pipet. For experiments in which the cells are to be partially activated by ACTH, the hormone is generally added to the cell suspension before distribution of the suspension among the individual incubation mixtures (e.g., 0.01 ml of ACTH solution, 25 mU/ml, is added per milliliter of suspension, in order to attain a final concentration of 0.1 mU/ml in the incubation medium). Each series of incubation mixtures includes a reagent blank (buffer only), a cell blank (unstimulated cells), and a corticosterone standard (0.5 or 1.0 μg/ml in buffer). The incubation mixtures are placed in a covered New Brunswick Gyrotory Shaker supplied with circulating 95% O_2-5% CO_2, and are allowed to incubate for 2 hr at 37°C. Optimal steroidogenic activity occurs in the absence of shaking.

B. Determination of Corticosterone

Corticosterone concentrations in the incubation medium are determined by a modification of the fluorometric method of Mattingly [18]. The two reagents required for the determination are dichloromethane (Merck Reagent, "for spectrophotometric use") and ethanol-sulfuric acid fluorescence reagent.

The latter is prepared by slow addition of 7 vol of concentrated, reagent grade sulfuric acid to 3 vol of redistilled absolute ethanol; and ice bath is used to absorb the heat generated during mixing.

After incubation of the adrenal cells, a 2-ml aliquot from each flask is transferred to a 40-ml glass-stoppered centrifuge tube containing 15 ml of dichloromethane. The mixture is shaken for 1 min, and the phases are separated by centrifugation for 1 min at 300 x g. The upper, aqueous phase is removed by aspiration and 19 ml of the lower, organic phase are added to 5 ml of ethanol-sulfuric acid reagent in a 15-ml glass-stoppered centrifuge tube and shaken for 20 sec. The upper, organic phase is aspirated and discarded. After an interval of 30-50 min to allow development of maximum intensity, the fluorescence of the lower phase is measured at excitation and emission wavelengths of 470 and 525 nm, respectively. For optimum sensitivity, use of a spectrofluorometer such as the Aminco-Bowman or Farrand Mark I is recommended. Fluorescence intensities are measured with the instrument meter set at zero with the reagent blank. Corticosterone concentrations are directly proportional to fluorescence intensity over the concentration range encountered in the incubation mixtures (up to 1.5 μg of corticosterone per ml);

concentrations of the steroid may, therefore, be calculated by comparison of the fluorescence intensities of unknowns with that of the corticosterone standard. In experiments in which agents are added to incubation mixtures in order to detect potential effects on steroidogenic activity of the cells, it is necessary that corresponding concentrations of the agents be tested as components of corticosterone standards to eliminate the possibility of direct interference through self-fluorescence or quenching by the agents.

C. Properties of Isolated Adrenal Cells

Adrenal cells incubated in the absence of stimulatory agents (cell blanks) display very low rates of apparent corticosterone secretion. These basal rates do not exceed 0.02 μg/ml/2hr, and generally fall in the range of 0.5-3.0% of the maximum rates of ACTH-activated steroidogenesis for the same cell preparations. Adrenal quarters, in contrast, typically demonstrate 4- to 8-fold stimulations during incubation with maximally stimulating concentrations of ACTH [8]. Inasmuch as a typical basal secretion rate for corticosterone from adrenal quarters is from 0.5 to 1.0 μg/adrenal/2 hr, the greater-fold response of isolated adrenal cells to ACTH would appear to reflect a lower basal steroidogenic rate of the cells relative to that of the adrenal quarters. The maximum steroidogenic rate for adrenal cells and quarters alike is about 4-8 μg/adrenal/2 hr [7,8,10-13,15].

The steroidogenic response of isolated adrenal cells to ACTH is illustrated in Figure 2. Threshold,

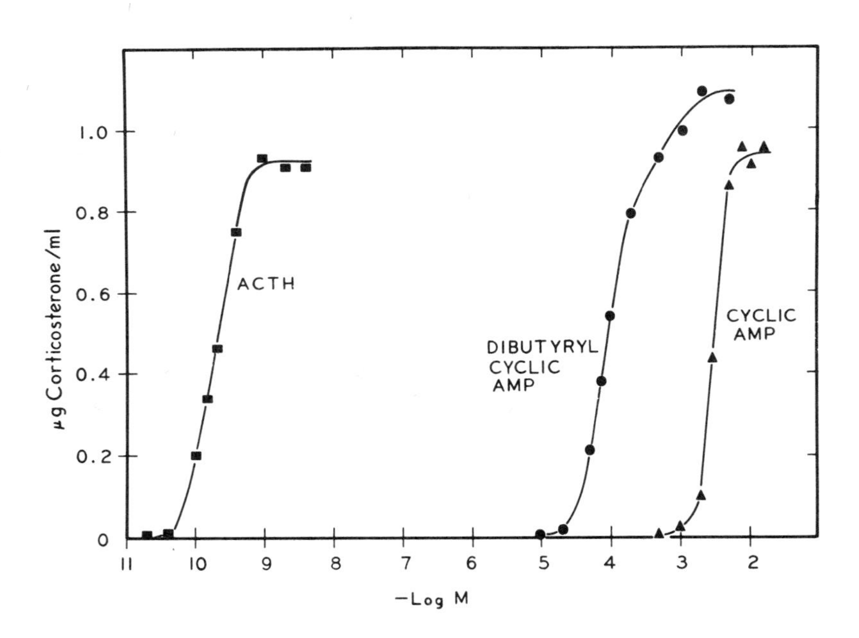

Fig. 2. Corticosterone secretion by adrenal cells in response to ACTH, dibutyryl cyclic AMP, and cyclic AMP. All points represent individual incubations within a single experiment. (Reprinted from Ref. [7], p. 3786, by permission of the American Chemical Society

half-maximal, and maximal responses to the hormone occur at concentrations of 0.04, 0.2, and 1 nM, respectively. Comparable degrees of sensitivity to ACTH have been reported for other isolated adrenal cell preparations [8,10-13,16], with threshold and half-maximal responses to the hormone occurring in the ranges of 0.002-0.2 and 0.2-2 nM, respectively. Two recent modifications [14,15] that use tryptic digestions and plastic or Teflon ware, however, have markedly increased the sensitivity of the adrenal cells to low concentrations of ACTH. The modification of Sayers et al. [15], for example, which also incorporates an increased concentration of calcium in the cell incubation medium, yields cells that display a threshold steroidogenic response at ACTH concentrations as low as 0.0003 nM (0.15 μU/ml).

The sensitivity of isolated adrenal cells to ACTH is markedly greater than that of intact adrenal tissue *in vitro*. Adrenal quarters, for example, demonstrate a steroidogenic response to ACTH concentrations over a range of about 20-2000 nM [8,19]. The lower sensitivity of the quarters is most likely owing to the inaccessibility of interior adrenal cells to ACTH in the absence of functional vascular apparatus within the tissue. The adrenal gland *in vivo*, on the other hand, demonstrates sensitivity to ACTH similar to that of the isolated adrenal cell. The intact adrenals of hypophysectomized 200-g rats, for example, secreted corticosterone at a threshold intravenous dose of 0.2 mU ACTH [20], an amount corresponding to an initial plasma concentration of roughly 0.05 nM.

Concentration-response curves for cyclic AMP and dibutyryl cyclic AMP, as well as for ACTH, are shown in Figure 2; half-maximally activating concentrations for the two nucleotides are about 3000 and 100 μM, respectively. Although the effective concentration ranges for ACTH and each of the nucleotides are considerably different, all three agents activate the cells to the same maximum steroidogenic rate. The latter observation is in accord with the role of cyclic AMP as a mediator, or "second messenger," in the steroidogenic action of ACTH [19].

IV. APPLICATIONS

Lipolytic and steroidogenic activities of the isolated cells are, in many instances, altered by agents that act directly at sites of synthesis, action, or degradation of cyclic AMP. Effectors that can increase the rates of lipolysis and steroidogenesis, for example, include activators of adenylate cyclase, inhibitors of cyclic AMP phosphodiesterase, and analogs of cyclic AMP that mimic the action of the nucleotide, whereas cyclic AMP-mediated cellular activities may be decreased by agents that inhibit adenylate cyclase. Examples of each of these four effector classes are presented here. Decreased cellular activity would also be expected in the presence of phosphodiesterase stimulators or compounds that inhibit at the site of action of cyclic AMP, although no examples of the latter two classes of agent are provided.

Table I includes examples from two effector classes that increase lipolytic and steroidogenic activity: (1) hormonal activators of adenylate cyclase, exemplified by ACTH and epinephrine; and (2) agents that mimic the action of cyclic AMP, as exemplified by a group of 3',5'-cyclic nucleotides. Both classes are characterized by the ability to produce maximum activation of lipolysis or steroidogenesis when added to otherwise unstimulated cells. In both cell types, the hormones and cyclic nucleotides have been observed to produce concentration-response curves having the same activity maxima or plateaus [7]. This common maximal response allows each activator to be characterized by an "A_{50}"

TABLE I

Activators of Lipolysis and Steroidogenesis

Agent	Lipocyte A_{50},μM	Adrenal A_{50},μM
Hormones		
ACTH	0.013	0.0002
Epinephrine	0.22	Inactive
Cyclic Nucleotides		
Cyclic AMP	8500	3300
Dibutyryl cyclic AMP	500	95
8-Methylthio cyclic AMP	180	65
8-Hydroxy cyclic AMP	260	90
8-Dimethylamino cyclic AMP	3000	130
6-Benzylthio analog of cyclic AMP	110	620

concentration, defined as the interpolated concentration required to produce half-maximal activation of lipolytic or steroidogenic activity. The activities of the cyclic nucleotides included in Table I appear to result from their ability to function as alternate activators of cyclic AMP-dependent protein kinases [7,21]. Cyclic AMP itself displays relatively low lipolytic and steroidogenic potency, a situation that may stem from poor ability of the nucleotide to penetrate cellular membranes. Although the cyclic nucleotides tend to be more potent activators of the adrenal cell, dramatic examples of selectivity are apparent. The 8-dimethylamino derivative of cyclic AMP, for example, is more than 20 times as potent in the adrenal cell as in the lipocyte, whereas the 6-benzylthio analog of cyclic AMP is roughly 5 times more potent a lipolytic than a steroidogenic agent.

The class of agents that stimulates lipolytic or steroidogenic activity through inhibition of cyclic AMP phosphodiesterase is exemplified by the four agents shown in the upper portion of Table II. Although these agents potentiate the effects of active hormones or cyclic nucleotides on the cellular activities, they generally do not, unlike the agents shown in Table I, stimulate cells in the unactivated, or "basal", state. For this reason, it is necessary to examine the effects of phosphodiesterase inhibitors in lipocytes or adrenal cells that have been partially activated by the addition of hormones or cyclic AMP. Epinephrine and ACTH concentrations sufficient to produce 25-50% of the maximal lipolytic or steroidogenic response, as used in Table II, are recommended for such experiments. Theophylline, SQ 20,006, and SQ 20,009 [the latter two compounds, derivatives of pyrazolo (3,4-b)-pyridines with potent

TABLE II

Effectors of Hormone-Activated[a] Lipolysis and Steroidogenesis

Effector	Lipolysis		Steroidogenesis	
	% of control	I_{50} (μM)	% of control	I_{50} (μM)
Stimulators				
Theophylline	245[b]	--	138[c]	--
SQ 20,006	300[b]	--	110[c]	--
SQ 20,009	172[b]	--	166[c]	--
Papaverine	159[b]	--	28[c]	--
Inhibitors				
Pyrvinium pamoate	--	0.02	--	0.15
Thioridazine	--	20	--	5
Puromycin	87[b]	--	--	17
Aminoglutethimide	--	--	--	30
Propranolol	--	0.10	89[b]	--

[a]Lipolysis and steroidogenesis were activated by 120 nM epinephrine and 0.15-0.20 nM ACTH, respectively.
[b]Effector concentration = 100 μM.
[c]Effector concentration = 200 μM.

activity as inhibitors of cyclic nucleotide phosphodiesterase, are described in Chapter 6, Section I.C] stimulate both lipolysis and steroidogenesis, although the orders of potency of the compounds differ in the two cell types. A fourth inhibitor of phosphodiesterase, papaverine [22], stimulates lipolysis, but inhibits steroidogenic activity. The inhibitory effect of papaverine in the adrenal cell is a typical response, reflecting inhibition at an additional site(s) in the cell.

The lower part of Table II provides examples of potent inhibitors of lipolytic or steroidogenic activity. The effects of these agents, like those of the phosphodiesterase inhibitors, are best determined in the presence of submaximally activating hormone concentrations, such as those indicated in Table II. In experiments with randomly selected compounds at concentrations of 100 μM, partial inhibition of lipolysis or steroidogenesis is frequently observed. Agents that inhibit by more than 50% at 100 μM are considered to be at least moderately potent inhibitors, and may be further characterized in terms of the concentration (I_{50}) required for 50% inhibition. Pyrvinium pamoate, an anthelmintic drug, is the most potent inhibitor yet found in either cell type. Puromycin and aminoglutethimide are relatively potent inhibitors of steroidogenesis, whereas propranolol, a β-adrenergic blocking agent, selectively inhibits epinephrine-activated lipolysis.

Agents that inhibit hormone-activated lipolysis or steroidogenesis through inhibition of adenylate cyclase may be detected by examining their effects on cyclic AMP-activated lipolysis or steroidogenesis. Table III illustrates this approach, as applied to

TABLE III

Effects of Inhibitors on Cyclic AMP-Activated[a] Lipolysis and Steroidogenesis

Inhibitor	Concentration (μM)	Lipolysis % of control	Steroidogenesis % of control
Thioridazine	10	90	106
	20	90	---
Propranolol	50	---	97
	100	100	92
Puromycin	20	---	46
	100	100	---
Aminoglutethimide	40	---	28
Pyrvinium pamoate	0.1	50	19
	0.2	---	13

[a]Lipolysis and steroidogenesis were activated by cyclic AMP at concentrations of 8 and 3 mM, respectively.

the inhibitors shown in Table II. The concentrations of cyclic AMP used are those producing roughly half-maximal activation of lipolysis or steroidogenesis, whereas the inhibitor concentrations are selected to approximate the I_{50} concentrations for the hormonally activated processes. Thioridazine, which does not significantly inhibit cyclic AMP-activated lipolysis or steroidogenesis, is an example of an inhibitor acting at the locus of adenylate cyclase in the cells, a site of inhibition consistent with the reported activity [23] of phenothiazines as inhibitors of hormonally activated adenylate cyclase in various subcellular preparations of the enzyme. The site of inhibition by propranolol may be similarly identified as the adenylate cyclase locus in the lipocyte, a conclusion consistent with the known property of propranolol as an inhibitor of epinephrine-activated adenylate cyclase prepared from fat cells [5].

The remaining compounds in Table III are equally effective at inhibiting hormone- or cyclic AMP-activated lipolysis or steroidogenesis, and are thus considered to inhibit at loci other than adenylate cyclase in the cells. For puromycin, a probable inhibitor of the synthesis of protein required for corticosteroidogenesis [24,25], and aminoglutethimide, an inhibitor of cholesterol desmolase [26], these sites of inhibition appear to be well identified. Pyrvinium pamoate, in contrast, provides an example of a powerful inhibitor with an as yet undetermined site of action in the lipocyte and adrenal cell.

ACKNOWLEDGMENTS

For those portions of the chapter concerning the lipocyte, many of the techniques and experimental data were generously provided by Dr. Mark Chasin. The excellent experimental assistance of Mrs. Velma Paik, Mr. Michael Rispoli, and Mr. Henry Sutter, as well as advice and encouragement from Dr. Sidney M. Hess, are also gratefully acknowledged.

REFERENCES

[1] G. A. Robison, R. W. Butcher, and E. W. Sutherland, Cyclic AMP, Academic Press, New York, 1971.

[2] M. Rodbell, J. Biol. Chem., 239, 375 (1964).

[3] H. F. DeLuca and P. P. Cohen, in Manometric Techniques, 4th ed. (W. W. Umbreit, R. H. Burris, and J. F. Stauffer, eds.), Burgess, Minneapolis, 1964, p. 13

[4] H. Ko and M. E. Royer, Anal. Biochem., 26, 18 (196

[5] L. Birnbaumer and M. Rodbell, J. Biol. Chem., 244, 3477 (1969).

[6] R. W. Butcher, C. E. Baird, and E. W. Sutherland, J. Biol. Chem., 243, 1705 (1968).

[7] C. A. Free, M. Chasin, V. S. Paik, and S. M. Hess, Biochemistry, 10, 3785 (1971).

[8] P. W. C. Kloppenborg, D. P. Island, G. W. Liddle, A. M. Michelakis, and W. E. Nicholson, Endocrinology, 82, 1053 (1968).

[9] I. D. K. Halkerston and M. Feinstein, Federation Proc., 27, 626 (1968).

[10] M. C. Richardson and D. Schulster, Biochem.J., 120, 25P (1970).

[11] R. Haning, S. A. S. Tait, and J. F. Tait, Endocrinology, 87, 1147 (1970).

[12] I. Rivkin and M. Chasin, Endocrinology, 88, 664 (1971).

[13] R. L. Swallow and G. Sayers, Proc. Soc. Exptl. Biol. Med., 13, 1 (1969).

[14] A. E. Kitabchi, R. K. Sharma, and W. H. West, Hormones Metab. Res., 3, 133 (1971).

[15] G. Sayers, R. L. Swallow, and N. D. Giordano, Endocrinology, 88, 1063 (1971).

[16] M. Nakamura and A. Tanaka, Endocrinol. Japon., 18, 291 (1971).

[17] D. R. Bangham, M. V. Mussett, and M. P. Stack-Dunne, Bull. World Health Organ., 27, 395 (1962).

[18] D. Mattingly, J. Clin. Pathol., 15, 374 (1962).

[19] D. G. Grahame-Smith, R. W. Butcher, R. L. Ney, and E. W. Sutherland, J. Biol. Chem., 242, 5535 (1967).

[20] G. W. Liddle, D. Island, and C. K. Meador, Recent Progr. Hormone Res., 18, 125 (1962).

[21] K. Muneyama, R. J. Bauer, D. A. Shuman, R. K. Robbins, and L. N. Simon, Biochemistry, 10, 2390 (1971).

[22] W. R. Kukovetz and G. Pöch, Arch. Exptl. Pathol. Pharmakol., 267, 189 (1970).

[23] J. Wolff and A. B. Jones, Proc. Natl. Acad. Sci. U. S., 65, 454 (1970).

[24] J. J. Ferguson, J. Biol. Chem., 238, 2754 (1963).

[25] L. D. Garren, W. W. Davis, R. M. Crocco, and R. L. Ney, Science, 152, 1386 (1966).

[26] M. P. Cohen, Proc. Soc. Exptl. Biol. Med., 127, 1086 (1968).

Chapter 9

ACCUMULATION OF CYCLIC AMP IN TISSUE SLICES AND INTACT CELLS: PRELABELING OF INTRACELLULAR POOLS OF ATP

John W. Daly

National Institute of Arthritis and Metabolic Diseases
National Institutes of Health
Bethesda, Maryland 20014

I. ACCUMULATION OF CYCLIC AMP FROM PRELABELED POOLS OF ADENINE NUCLEOTIDES IN INTACT CELLS

The concentrations of cyclic AMP in intact cells reflect both the rate of synthesis and the rate of degradation of this substance. Synthesis is catalyzed by membrane adenylate cyclases, whose activity is controlled by extracellular hormones through interact-ions with membranal receptors. The rate of synthesis will depend also upon the local intracellular concentration of ATP. The degradation of cyclic AMP is catalyzed by specific phosphodiesterases.

Recently, radiometric techniques have been introduced for the study of the accumulation of cyclic AMP in intact cells. These techniques consist of prelabeling endogenous adenine nucleotides by incubati of tissue slices, cells, or perfused organs with radioactive adenine or adenosine. The radioactive intracellular ATP then serves as a substrate for adenylate cyclase, and is partially converted to radioactive cyclic AMP on stimulation of the preparation with appropriate hormones. After termination of the experiment, carrier cyclic AMP is added to the preparation, reisolated, and radiometrically assayed. The accumulation of radioactive cyclic AMP may be expressed in terms of levels of radioactivity in the biological preparation, or in terms of percent of total radioactivity in the preparation that has been converted to cyclic AMP. Purification of the accumulated radioactive cyclic AMP is of the utmost importance, since the ultimate assay is nonspecific and is based solely on radioactivity associated with the carrier cyclic AMP. A significant amount of radioactivity in the biological preparation may be incorporated also into RNA and DNA.

Incorporation of adenine into the intracellular nucleotides presumably involves passive diffusion of the adenine across the cell membrane, followed by conversion to 5'-AMP by the action of adenine phosphoribosyltransferase with 5-phosphoribosyl pyrophosphate as the cosubstrate [1]. The rate of formation of 5-phosphoribosyl pyrophosphate might limit the incorporation of adenine into intracellular nucleotides in certain tissues.

Incorporation of adenosine into intracellular nucleotides may also involve an initial passive diffusion across the membrane, followed by conversion of intracellular adenosine to 5'-AMP by the action of adenosine kinase, with ATP as the cosubstrate and ADP as the other product [2]. The uptake of adenosine by mammalian cells may not, however, be an entirely passive process [3]. An alternative route for incorporation of adenosine into intracellular nucleotides has been demonstrated in *E. coli*, where adenosine is first converted to adenine by the action of adenosine phosphorylase, then the adenine is converted to 5'-AMP by adenine phosphoribosyltransfer [4].

Intracellular adenine nucleotides may be labeled also by incubation of preparations with radioactive hypoxanthine or inosine. In the case of hypoxanthine incorporation would depend on its conversion to 5'-IMP by the action of hypoxanthine phosphoribosyltransferase, followed by conversion to 5'-AMP by the action of the enzymes adenylosuccinic synthetase and adenylosuccinase. Incorporation of inosine may follow one of two pathways, either direct conversion to 5'-IMP by the action of inosine kinase or phosphorolysis to hypoxanthine by the action of inosine phosphorylase, followed by incorporation of hypoxanthine as described.

Labeling of nucleotides in intact cells with adenine or adenosine depends on the activity of adenine phosphoribosyltransferase and adenosine kinase, respectively. Whether or not all of the adenine nucleotides within the cell are uniformly labeled depends on the intracellular distribution of these enzymes and on whether or not the adenine nucleotides within cellular compartments, such as mitochondria and cell plasma, equilibrate rapidly. Adenine phosphoribosyltransferase has been reported to be a soluble enzyme, while adenosine kinase appears to be partially associated with membrane fragments [5]. Results obtained with various preparations indicate that labeling procedures with adenine and adenosine do not result in a uniform distribution of specific activities among the adenine nucleotides within the cell. Thus, use of the prelabeling technique to measure accumulation of cyclic AMP must be used with caution. The specific activity of the ATP cannot be used *a priori* to calculate endogenous concentrations of cyclic AMP from the levels of radioactivity associated with recovered carrier cyclic AMP. Three situations may pertain, and the literature appears to contain examples of all three.

(1) The radioactive precursor may be incorporated mainly into adenine nucleotides that do not effectively serve as precursors for cyclic AMP.
(2) The radioactive precursor may be incorporated selectively into a pool of adenine nucleotides that serve as excellent precursors of cyclic AMP.
(3) The radioactive precursor may be incorporated uniformly into adenine nucleotides of the cell.

The additional possibility of translocation of adenine nucleotides between intracellular compartments during stimulated formation of cyclic AMP also must be considered. In spite of these problems, the technique of using prelabeled intact cells to study the accumulation of cyclic AMP can, especially when used in conjunction with determination of endogenous cyclic AMP, provide valuable information on the factors affecting formation and degradation of cyclic AMP in cells, tissues, and organs.

II. APPLICATION OF THE PRELABELING TECHNIQUE TO VARIOUS BIOLOGICAL PREPARATIONS

The technique of prelabeling intact cells with radioactive adenine or adenosine has now been used in many biological preparations to study the accumulation of cyclic AMP. The results have not been uniformly successful, perhaps owing to the levels and intrcellular distribution of the enzymes responsible for incorporation of adenine and/or adenosine in different preparations. Methodology and results obtained with the prelabeling technique in certain biological preparations are summarized in Table I, and are disccussed in the following sections.

TABLE I

Labeling of Adenine Nucleotides in Tissue Slices and Intact Cells with Radioactive Adenine and Adenosine: Accumulation of Intracellular Radioactive Cyclic AMP

Preparation [Ref.]	Precursor (μM)	Prelabeling Time (min)	Stimulation of Accumulation of Cyclic AMP
Fat pads [9]	[^{14}C]-Adenine (100-200)	120	NE,[a] 6-fold ACTH, 6-fold
Fat cells [11]	[^{14}C]-Adenine (20)	120	NE, 40-fold
Fat cells [12]	[^{3}H]-Adenine (0.4)	15	NE, 0-fold
Blood platelets [22]	[^{14}C]-Adenosine ---	120	PGE_1,[b] 6 to 7 fold
Blood platelets [24]	[^{14}C]-Adenine (2)	80	IPA,[c] 3-fold
Erythrocytes [28]	[^{3}H]-Adenosine (1)	120	IPA, 2-fold
Leukocytes [29.30]	[^{3}H]-Adenine (0.1)	40	H,[d] 2-fold PGE_1, 10-fold
Brain slices [36,37]	[^{14}C]-Adenine (1.3)	40	Various Agents, up to 100-fold
	[^{3}H]-Adenosine (1.3)	40	Various Agents, up to 50-fold
Heart slices [68]	[^{14}C]-Adenine ---	45	NE, 2-fold
Liver slices [68]	[^{14}C]-Adenine ---	45	Glu,[e] 2-fold
Testicular slices [69]	[^{14}C]-Adenine (10)	60	LH,[f] 10-fold

TABLE I (continued)

Preparation [Ref.]	Precursor (μM)	Prelabeling Time (min)	Stimulation of Accumulation of Cyclic AMP
Ovary [70]	[^{14}C]-Adenine (10)	60	LH, 30-fold
Anterior pituitary [71]	[^{3}H]-Adenine (3)	60	PGE_1, 7-fold

[a]NE = Norepinephrine
[b]PGE_1 = Prostaglandin E_1
[c]IPA = Isoproterenol
[d]H = Histamine
[e]Glu = Glucagon
[f]LH = Luteinizing Hormone

A. Isolated Cells

The use of isolated and relatively homogeneous cell populations greatly simplifies the interpretation of results obtained by the prelabeling technique. Thus, the problems of differential labeling of different cell types present in tissue slices and perfused organs do not pertain and the primary complication concerns intracellular compartmentalizati of labeled ATP and cyclic AMP. Uniformity of cell type and viability of the cells, however, are still major considerations.

1. Adipose Cells

The prelabeling technique was first applied to

the study of adipose cells isolated from labeled rat fat pads. After fat pads were labeled with radioactive adenine, increases in radioactive cyclic AMP could be demonstrated in dispersed adipose cells incubated with norepinephrine and other hormones [6-9]. The experimental procedure to label fat pads has been varied only slightly, and is essentially as follows.

Fat pads from 15 male Sprague-Dawley rats were incubated with shaking for 1.5 hr at 37°C in 4 ml of a Krebs-Ringer bicarbonate solution, pH 7.4, containing 40 μmoles glucose, 160 mg of bovine serum albumin, and 0.1 mM [^{14}C]-adenine (10 μCi); then 7 mg of collagenase and 10 μCi of additional [^{14}C]-adenine were added and incubation was continued for 40 min. (See Chapters 4 and 8 for further details.) The dispersed cells were then filtered, washed, and resuspended in Krebs-Ringer bicarbonate solution containing 4% bovine serum albumin to give a solution containing 40-50 mg of cells/ml. Approximately 17% of the [^{14}C]-adenine was incorporated into the cells [9], but apparently only 8-10% of this was present in the trichloroacetic acid (TCA)-soluble supernatant [6-8]. After incubation of 1 ml of dispersed cells in the presence of various agents, either the medium plus cells, medium alone, or cells alone were extracted with organic solvents to remove lipids, followed by precipitation of proteins with TCA. The [^{14}C]-cyclic AMP was measured in the aqueous solution after three $BaSO_4$ precipitations [10]. Corrections for recovery of cyclic AMP could be made by inclusion of either exogenous unlabeled or radioactive cyclic AMP. The purity of the [^{14}C]-cyclic AMP was ascertained by paper chromatography. Control incubations with theophylline contained about 5000 cpm of [^{14}C]-cyclic AMP per 100 mg of cells, while the presence of norepinephrine or corticotropin increased the concentration to nearly 30,000 cpm/100 mg of cells within 20 min. The latter value represented about 12% of the soluble adenine nucleotides present in the cells [9].

An alternative method [11] involves the labeling of isolated adipose cells by incubation with [^{14}C]-adenine as follows.

Rat adipose cells (see Chapters 4 and 8) equivalent to 200 mg of tissue were suspended in 1 ml of Krebs-Ringer phosphate buffer, pH 7.2, containing 2% bovine serum albumin, 6 mM glucose, and 0.02 mM [^{14}C]-adenine (1 μCi), and were incubated for 2 hr at 37°C with shaking. At this time, nearly complete incorporation of [^{14}C]-adenine into the cells had occurred. Agents were now added to the medium and incubation was continued until terminated by the adddition of TCA. Carrier cyclic AMP was added and the material was chromatographed on Dowex 50, then purified further by two $BaSO_4$ precipitations (described more fully in Chapter 4) and paper chromatography. This extensive purification was reported to be necessary to remove radioactive contaminants from the cyclic AMP. Recoveries of carrier cyclic AMP were about 40%. It is noteworthy that the authors [11] state that although incorporation of [^{14}C]-adenine into ATP was maximal in 40 min, maximal conversions to cyclic AMP were attained only after an additional 80 min of preincubation. Control cells incubated in the presence of theophylline for 5 min contained radioactivity equivalent to 400 cpm of [^{14}C]-cyclic AMP per 100 mg of original tissue. This level increased in the presence of theophylline and norepinephrine to 16,000 cpm/100 mg of original tissue. The accumulation of [^{14}C]-cyclic AMP in the presence of norepinephrine and theophylline, in contrast to the results of Kuo and De Renzo [9] was maximal in less than 5 min. Counting efficiencies were not reported [11], so it is impossible to calculate accurately what percent of the total adenine nucleotides are present in [^{14}C]-cyclic AMP. If we assume a counting efficiency of more than 20% for ^{14}C, the conversion with norepinephrine and theophylline would be less than 5%.

The use of short labeling times and low concentrations of radioactive adenine with isolated adipose cells greatly increased the norepinephrine-evoked conversion of labeled nucleotides to cyclic AMP [12,13]. The methodology is as follows.

Isolated rat adipose cells were incubated for 15 min at 37°C with 4 μmoles of [^{3}H]-adenine (10 μCi) in 4 ml of Krebs-Ringer phosphate buffer, pH 7.4, containing 5% albumin. The cells were then washed and suspended in 50 vol of Krebs-Ringer phosphate buffer containing 5% albumin. After an additional 5 min in the presence of various agents, the incubations were terminated by the addition of 1.3% perchloric acid and carrier cyclic AMP. Cyclic AMP was purified by Dowex-50 chromatography, followed by two $BaSO_4$ precipitations. Results were corrected for recovery of carrier cyclic AMP. The purity (90-100%) of the [^{3}H]-cyclic AMP was ascertained by paper chromatography.

Under these labeling conditions of short time and low adenine concentrations, ATP, ADP, and AMP represented 95% of the radioactivity in the cells. In the presence of norepinephrine and theophylline, nearly 30% of the total radioactive adenine nucleotides could be converted to cyclic AMP. Conversions were maximal within 5 min.

The extreme variation in the results obtained with rat adipose cells labeled with radioactive adenine under different conditions, then incubated with norepinephrine under similar conditions, provides strong evidence for the existence of compartmentalized pools of ATP in such cells. Studies of the relationship of lipolysis induced by dibutyryl cyclic AMP to the concentrations of ATP in adipose cells also provided evidence for intracellular compartmentalization and function of ATP pools [14]. Further studies in which the specific activities of cyclic AMP, ATP, ADP, and AMP in adipose cells are measured after prior labeling with adenine or adenosine could provide insights into the nature of this intracellular compartmentalization.

2. Ascites Tumor Cells

Incorporation of [^{14}C]-adenosine into RNA, DNA, and soluble adenine nucleotides has been studied with mouse ascites tumor cells [15]. It was noted that the specific activity of ATP and ADP was much higher than that of 5'-AMP, evidence for intracellular compartmentalization of adenine nucleotides in these cells. Incorporation into cyclic AMP was not determined.

3. Glioma Cells

The accumulation of radioactive and endogenous cyclic AMP in glioma cells has been compared after the cells were labeled with adenine or adenosine [16].

Cells were incubated for 40 min with 4 μM [^{14}C]-adenine (2 μCi) in 10 ml of Krebs-Ringer bicarbonate solution aerated before incubation with 95% O_2 - 5% CO_2. Incubations in fresh medium with various agents were terminated by the addition of TCA. Radioactive cyclic AMP was isolated by thin-layer chromatography (see Section II.B.1). Both radioactive adenine and adenosine were effectively converted to adenine nucleotides. Under control conditions, cyclic AMP represented less than 1% of the total intracellular radioactivity. The concentrations of radioactive cyclic AMP were increased greatly in a number of cell lines derived from nervous tissue upon incubation with norepinephrine. Endogenous levels of cyclic AMP were also determined by use of a protein-binding assay [16] (see Chapter 2). The results indicated that adenine nucleotides are compartmentalized in glioma cells, and that incubation with [^{14}C]-adenine labels adenine nucleotides that serve as precursors for cyclic AMP more specifically than incubation with [^{14}C]-adenosine.

4. Blood Platelets

The [^{14}C]-adenosine is incorporated into ATP, ADP, and 5'-AMP of human blood platelets [18-20]. The major portion of its radioactivity is found in ATP and ADP. The specific activity of the ATP, however, was consistently 3 to 5 times that of ADP, suggesting that a large portion of the intracellular ADP is metabolically inert. Adenine nucleotides of blood platelets can also be labeled by incubation with radioactive adenine [21]. Adenosine inhibits incorporation of adenine, but the converse is not true [21].

Prelabeling of intracellular ATP in human platelets by incubations with [^{14}C]-adenosine has been used to investigate the subsequent accumulation of cyclic AMP as follows [22].

Human platelet pellets were suspended in 0.04 M tris-HCl-salt medium, pH 7.4, containing 6.7 mM EDTA, and incubated for 2 hr at 37°C with [^{14}C]-adenosine

[20 Ci/mole). Aliquots were then incubated in medium containing 50 mM caffeine. The incubation was terminated by heating at 100°C for 5 min after the addition of carrier cyclic AMP. Cyclic AMP was isolated by the method of Krishna et al. [23]. Prostaglandin E_1 caused an increase of 6 to 7-fold in the accumulation of [^{14}C]-cyclic AMP.

The formation of [^{3}H]-cyclic AMP from platelet ATP prelabeled by incubation of platelet-rich plasma with [^{3}H]-adenine has been reported [24] as follows.

Human platelet-rich plasma (4 x 10^{8} platelets/ml) was incubated 80 min at 37°C with 2 μM [^{14}C]-adenine (0.6 μCi). At this time, 85% of the radioactivity was incorporated into platelets. Test substances were added; further incubations were terminated by the addition of perchloric acid containing [^{3}H]-cyclic AMP to determine recoveries. Purification of cyclic AMP was by ion-exchange chromatography, followed by thin-layer chromatography with carrier cyclic AMP. After control incubations of 1 min with the phosphodiesterase inhibitor, papaverine, [^{14}C]-cyclic AMP represented 0.06% of the total radioacitivity in the slice. Isoproterenol increased this level 3-fold. Prostaglandin in the presence of theophylline or caffeine also increased the accumulation of radioactive cyclic AMP in platelets prelabeled with adenine [25,26]

Compartmentilization of adenine nucleotides in platelets was inferred from studies on the relationship of aggregation, inhibition of ATP formation, and incorporation of radioactive adenosine and adenine [20]. It was postulated that adenosine kinase activity reduces ATP concentrations necessary for the energy-requiring process of aggregation, while synthesis of phosphoribosyl pyrophosphate for the incorporation of adenine does not.

5. Erythrocytes

Rat erythrocytes prelabeled by incubation with [^{14}C]-adenine contain 96% of their radioactivity as adenine nucleotides, and release radioactive hypoxanthine on further incubation [27]. Frog or tadpole erythrocytes labeled with [^{3}H]-adenosine contain 85-90% of intracellular radioactivity present as ATP [28]. Tadpole erythrocytes incorporate much more radioactivity than do frog erythrocytes. Labeling and assay for [^{3}H]-cyclic AMP accumulation in amphibian erythrocytes [28] were conducted as follows.

Erythrocytes (0.5 ml packed cell volume) were suspended in 5 ml of amphibian Ringers solution, and were incubated with shaking for 2 hr at 30°C. They were then centrifuged, resuspended, and incubated for

2 hr in 2 ml of Ringer's solution containing 10 mM caffeine and 1 μM [^{3}H]-adenosine (50 μCi). Isolated cells were washed three times, resuspended in 2 ml of the Ringer's solution, and incubated with test agents for 1 hr at 30°C. The cells were isolated and lysed with H_2O; the solution was precipitated with $BaSO_4$. Carrier cyclic AMP was added, and reisolated by paper chromatography. The original low concentrations of [^{3}H]-cyclic AMP were increased twofold in frog erythrocytes on incubation with isoproterenol and caffeine. No increase was observed with tadpole erythrocytes, probably owing to the lack of a functional catecholamine receptor in the tadpole cells [28].

6. Leukocytes

The [^{3}H]-adenine incorporated into adenine nucleotides in human leukocytes serves as a precursor of [^{3}H]-cyclic AMP [29,30]. Labeling and assay of accumulation of [^{3}H]-cyclic AMP were conducted essentially as follows:

Human leukocytes (20×10^7) were suspended in 20 ml of Hanks basic salts solution containing 25% human AB serum and 0.1 μM [^{3}H]-adenine (20 μCi), and were incubated for 40 min at 37°C. The labeled cells incorporated nearly 80% of the radioactive adenine;

they were then either centrifuged and resuspended in tris, or were used directly for further incubation with test agents. Cells were collected by centrifugation, resuspended with the addition of carrier cyclic AMP, and heated at 100°C. Isolation of cyclic AMP was by Dowex-50 chromatography, followed by $BaSO_4$ precipitation.

In control incubations with 10 mM theophylline, [^{3}H]-cyclic AMP represented about 0.3% of the total [^{3}H]-adenine incorporated. This concentration of [^{3}H]-cyclic AMP was increased twofold by incubation with histamine [29] and up to 10-fold by incubation with prostaglandin E_1 [30]. Similar accumulations of [^{3}H]-cyclic AMP were elicited by prostaglandin E_1 in labeled preparations of human lymphocytes [30].

B. Tissue Slices

Investigation of the accumulation of radioactive cyclic AMP with prelabeled tissue slices is complicated by the possibility of differential labeling of different cell types, in addition to the aforementioned intracellular compartmentilization of labeled adenine nucleotide. The method has, however, been extensively applied in such preparations and may possibly provide valuable information on sites of generation of cyclic

AMP in such heterogenous cell populations.

1. Brain

Procedures for the preparation and incubation of viable brain slices for long periods of time (2-3 hr) have been extensively studied [31,32]. Such incubated slices incorporate adenine, adenosine,inosine, and hypoxanthine into intracellular adenine nucleotides, RNA, and DNA [33,34]. Adenine and adenosine are excellent precursors of adenine nucleotides, with about 85-90% of the radioactivity incorporated from these precursors present in rat cerebral cortical slices as adenine nucleotides. Hypoxanthine and inosine are much less effective and specific as nucleotide precursors. The uptake of adenine appears to be dependent on the activity of adenine phosphoribosyltransferase [1,33], while the incorporation of adenosine is apparently catalyzed by adenosine kinase [2]. Adenine is also taken up into rat brain and incorporated into adenine nucleotides *in vivo* [35].

The labeled adenine nucleotides derived from radioactive adenine in cerebral cortical and cerebellar slices served as excellent precursors of radioactive cyclic AMP in the presence of various agents [36]. The percent accumulation of radioactive cyclic AMP from labeled nucleotides of cortical slices was much less (50% less) when the slices had been prelabeled with adenosine instead of adenine [36,37].

The use of slices prelabeled with radioactive adenine has been used to study the accumulation of cyclic AMP in brain tissue from various species [38-41]. Slices from cerebral cortex, cerebellum, caudate nucleus, and hypothalamus have been studied [38]. Human cortical [42,43] and cerebellar tissues [43] have been investigated by this technique. Cortical grey matter gave large responses to norepinephrine, while white matter was nonresponsive [42]. Accumulations of radioactive cyclic AMP in cerebral cortical and cerebellar slices prelabeled with adenine have been reported in response to (nor)epinephrine and related catecholamines [45], histamine [44,46], serotin and analogs [41,44], adenosine [37,39,47], depolarizing agents [47-49], tricyclic antidepressants [44,50], and tranquilizers [44], and various combinations of agents [39,41,44, 45,47,51]. Prostaglandin E_1 stimulated accumulations of radioactive cyclic AMP in rat cerebral cortical slices prelabeled with adenine [52]. Prelabeled slices have been used also to study the release of radioactive adenosine in response to depolarizing agents [41], electrical pulsation [53], and tricyclic pyschotropic agents [54].

Adenosine appears to be involved in the accumulation of radioactive cyclic AMP elicited under such conditions. The observed increases of radioactive cyclic AMP in prelabeled slices are quite similar to the increases in endogenous levels of cyclic AMP reported by Kakiuchi and Rall [55,56]. Technical difficulties, which are apparently owing to the use of relatively thick brain slices (1 mm), have prevented successful application of the prelabeling technique to the study of electrically-pulsed slices [57]. Electrical pulses do increase endogenous levels of cyclic AMP [58].

A satisfactory method [36] to label brain slices of cortical tissue with adenine or adenosine is as follows.

Male Hartley strain guinea pigs of about 300 g are stunned and killed by severence of the carotid artery and trachea. The brains are immediately removed and immersed in ice-cold Krebs-Ringer glucose solution [36]. The gray matter of the cerebral cortex is cut (by hand) with a razor blade into thin slices (usually some 25 slices are cut from one brain) and immediately transferred into ice-cold Krebs-Ringer bicarbonate-glucose solution, which has been gassed with 5% CO_2-95% O_2. The tissue from one brain

(0.8-1 g wet weight) is then placed in a single layer on the platform of a McIlwain Mechanical Tissue Chopper (Brinkmann Instruments, Inc., Westbury, N.Y.), blotted slightly to remove excess liquid, and sliced at room temperature with the blade adjustment set at 0.25 mm. The platform is topped with a thin plastic disc and two pieces of filter paper, and finally covered with aluminum foil for this operation.

The slices are now quickly transferred for preincubation into a 30-ml beaker containing 18 ml of Krebs-Ringer bicarbonate-glucose solution and incubated at 37°C for 15 min. All incubation solutions are equilibrated at 37°C prior to the addition of the slices, and are gassed continuously with a mixture of 5% CO_2-95% O_2. After preincubation, the slices are collected on a small Buchner funnel with the aid of a water aspirator, then transferred with a spatula to a 30-ml beaker containing 10 ml of the Krebs-Ringer bicarbonate-glucose solution and 13 μM [^{14}C]-adenine or 13 μM [^{14}C]-adenosine (1.5 μCi). After cells are labeled at 37°C for 40 min, the incubation medium is decanted and the slices are washed twice with the Krebs-Ringer bicarbonate-glucose, equilibrated at 37°C. The labeled slices are collected on a small Buchner funnel and divided into 6-8 portions. Each portion is transferred with a spatula to a 15-ml beaker containing 6 ml of the Krebs-Ringer bicarbonate-glucose solution, and incubated for 10 min before the addition of the test substance or transfer to the Krebs-Ringer solution that contains test substances, such as 40-125 mM potassium ions. The incubation is terminated by collection of the slices on a Buchner funnel and the collected slices are transferred with a spatula into 1.0 ml of ice-cold 8% (v/v) TCA in a small ground-glass homogenizer, followed by immediate homogenization by hand. Carrier cyclic AMP (300 μg) in 0.3 ml of H_2O is added to the homogenate, which is centrifuged for 10 min at 300 x g. The supernatant solution is transferred to a 10-ml screw-cap tube. The total radioactivity in the supernatant fraction is determined by counting a 30-μl portion in a vial containing 12 ml of Bray's solution. The remaining portion of the supernatant is extracted once with an equal volume of benzene, and twice with equal volumes of ether to remove lipids and excess TCA, and evaporated to dryness at 50°C under a stream of nitrogen (30 min). The residue is dissolved in 150 μl of 50% aqueous ethanol for thin-layer chromatography on a 5 x 20 cm x 250 μm slicia gel GF plate with n-butyl alcohol-ethyl acetate-methanol-ammonium hydroxide (7:4:3:4) as solvent. Multiple development may be used to provide better separations of cyclic AMP from other compounds. The ultraviolet-absorbing band of carrier cyclic AMP is removed, and the cyclic AMP is extracted with 3 ml of 50% aqueous ethanol. The absorption at 258 nm of a 0.3-ml portion of the 50% aqueous ethanol extract is measured in 0.7 ml of H_2O to calculate the recovery of the carrier cyclic AMP. The radioactivity of a 2.0-ml portion is determined in 10 ml of Bray's solution.

Several modifications of this methodology are possible. For example, the medium may be removed by decantation through a 400-mesh nylon net supported on layers of absorbent tissue paper. This procedure allows collection of slices without undue drying.

The absolute values of percent conversion with various agents obtained with such prelabeled slices are apparently dependent on various unknown factors. This variation has been discussed [1]. Attempts to elucidate the responsible factors have not been completely successful. The degree of aeration and of agitation of slices appears important, especially with regard to the rate of cyclic AMP accumulation elicited by depolarizing agents. Maximal accumulation of cyclic AMP with most agents does not occur until after 4-15 min [36,49]. This slow rate of accumulation may, in part, reflect the rate of diffusion of the agent into the slice [59].

The length of the labeling period with adenine, and the condition of the slices during this period, can affect the subsequent accumulation of cyclic AMP. Thus, during the first minutes of labeling [60] and during labeling of slices in the presence of 40 mM potassium ions [61], adenine nucleotides are formed that are better precursors of cyclic AMP than those formed during the normal 40-min labeling period. Labeling for short periods of time in the presence of the elevated steady-state concentrations of endogenous cyclic AMP elicited by the presence of histamine results in enhanced selective incorporation of [^{3}H]-adenine into cyclic AMP [62]. In such slices, the specific activity of cyclic AMP was 7-fold that of the ATP in the presence of histamine, and only 2-fold that in the absence of histamine. Thus, variations in the concentrations of certain endogenous factors that stimulate cyclic AMP accumulation such as, for example, histamine, norepinephrine, or adenosine [63], could greatly influence the selective labeling of precursor pools of adenine nucleotides.

The concentration (6-13 μM) of adenine does not affect subsequent accumulation of radioactive cyclic AMP [37]. In three published modifications of the original method [36] incubations of 40 min with 5 nM [45], 20 μM [38,40], and 0.5 μM [43] adenine were used to label slices.

After slices are labeled for 40 min with adenine, the precursor pool of adenine nucleotides is remarkably stable after incubation, in the absence of a stimulant [41]. However, during several repetitive stimulations of radioactive cyclic AMP accumulation with histamine + norepinephrine + adenosine, the percent conversion of adenine nucleotides to radioactive cyclic AMP decreases [59]. Since the precursor pool is remarkably stable in the absence of stimulants, incubation for 10-30 min after prelabeling can, with cortical slices from rat and mouse, be used to reduce control concentrations of radioactive cyclic AMP to minimal values [38].

Prelabeling studies with adenine have been reported with either slices prepared in the cold and then incubated at 37°C [36,45], or with slices prepared at room remperature [38,40]; the results obtained appear to be comparable. The temperature of preparation of cortical slices, has however, been reported [62,64] to influence the effect of theophylline on the histamine-elicited accumulation of endogenous cyclic AMP in guinea pig cortical slices. It is noteworthy that phosphodiesterase activity in cortical slices does not appear to be readily inhibited by theophylline [59]. In addition, theophylline *blocks*, rather than potentiates, the accumulation of radioactive cyclic AMP elicited by various agents [44,47-50,54,59]. Although theophylline blocks the stimulatory effect of adenosine [63], it does not block the uptake or incorporation of adenosine into intracellular adenine nucleotides [36,63]. For inhibition of phosphodiesterase in brain slices, papaverine or 3-isobutyl-1-methyl-xanthine is much more effective than theophylline [59].

Radioactive cyclic AMP may be isolated from brain slices for assay by thin-layer chromatography [36], as described, or by the method of Krishna et al. [23], as in Refs. [38,40,43,45] (see also Chapter 4). Regardless of the method used, the radioactive purity of the isolated cyclic AMP is of great importance. In the thin-layer chromatographic separation, radioactive hypoxanthine, which migrates only slightly further than cyclic AMP, can contribute to the total radioactivity recovered with the carrier cyclic AMP [59].

For the measurement of the specific activity of $[^{14}C]$-cyclic AMP, a small amount of $[^{3}H]$-cyclic AMP of high specific activity may be added to ascertain recoveries in one aliquot, while endogenous concentrations are assayed in another aliquot [40,60]. In other approaches to this problem, either the recovery of cyclic AMP [62] or the endogenous levels of cyclic AMP [59] can be determined in parallel experiments. The specific activity of cyclic AMP was reported to be relatively constant after stimulation of cyclic AMP accumulation by various agents [60]. In contrast to these results, the rates of formation of radioactive cyclic AMP and endogenous cyclic AMP were not the same [40,59]. The reason for this apparent dichotomy is unknown.

The morphological location of the precursor pool or pools labeled by incubation of brain slices with radioactive adenine or adenosine is unresolved. Preliminary results of radioautography combined with electron microscopy suggest a neuronal localization for the majority of the radioactivity [65]. Glioma cells, however, respond to norepinephrine by the accumulation of cyclic AMP [16,66,67], and such cells contain active adenosine kinase and adenine phosphoribosyltransferase activity that incorporates radioactive adenine and adenosine into precursor pools of adenine nucleotides [16] (see Section I.A.3). It appears likely that the accumulation of radioactive cyclic AMP in labeled brain tissue reflects participation of both neuronal and glial elements.

2. Heart

Rat heart ventricular slices incorporate [^{14}C]-adenine into intracellular adenine nucleotides [68]. The procedure for prelabeling and investigation of the accumulation of [^{14}C]-cyclic AMP is as follows:

Slices (50-75 mg) are incubated with shaking for 45 min at 37°C in an atmosphere of 95% O_2-5% CO_2 in 2 ml of a Krebs-Ringer bicarbonate buffer containing 1.4 mM calcium ions, 2 mg of glucose, 2 mg of albumin, and [^{14}C]-adenine (2 μCi). The medium is then replaced with fresh medium, and test agents are added after 5 min of further incubation. Incubations are terminated by boiling after the addition of carrier cyclic AMP and [^{3}H]-cyclic AMP to determine recoveries. Cyclic AMP is isolated by paper chromatography. Under control conditions, 150-200 cpm of [^{14}C]-cyclic AMP is present per 100 mg of tissue. This concentration is doubled by the presence of caffeine and norepinephrine.

3. Liver

Liver slices are prepared and incubated with [^{14}C]-adenine as described above for heart [68]. After control incubations of 5-min duration with caffeine, about 100 cpm/mg of tissue are present as [^{14}C]-cyclic AMP. These concentrations are doubled by glucagon.

4. Testis

Rat testes are isolated, cut into small sections, and labeled [69] as follows.

The tissue (60-100 mg) is incubated for 1 hr at 37°C, with shaking, in 2 ml of Krebs-Ringer phosphate buffer, pH 7.2 containing 1.4 mM calcium ions, 6 mM glucose, and 0.01 mM [^{14}C]-adenine (1 μCi). Agents and 5 mM theophylline are then added, and incubations are continued for 30 min, followed by homogenization with carrier ATP, ADP, 5'-AMP, cyclic AMP, adenosine, and adenine. Cyclic AMP is isolated by a method described for adipose cells [11]. It is stated that the radioactive adenine requires purification by ion-exhange chromatography before use. After control incubations, 4000 cpm of cyclic AMP are present per 100 mg of tissue. This level is increased nearly 10-fold by incubations with luteinizing hormone (LH) and follicle-stimulating hormone (FSH). Interpretation of the results is complicated by the apparent contamination of FSH preparations by LH. However, in young rats and hypophysectemized rats the response to FSH is more than twice that to LH. Similar responses to FSH and LH are reported in isolated seminiferous tubules [69].

5. Ovary

Intact mouse ovaries are labeled by incubation for 40 min with 0.01 mM [^{14}C]-adenine and the accumulation of radioactive cyclic AMP is measured after a further 30-min incubation with test agents and 5 mM theophylline [70]. The method is essentially as described for testis [69]. Luteinizing hormone causes an increase to 30-fold in [^{14}C]-cyclic AMP content, while prostaglandin E_1 causes an increase to 60-fold. Control levels of radioactive cyclic AMP are about 700 cpm/100 mg tissue.

6. Anterior Pituitary Gland

Whole rat anterior pituitary glands are incubated for 1 hr at 37°C in 2 ml of Krebs-Ringer bicarbonate buffer containing 2 mg glucose and 0.25 μM [^{3}H]-adenine (10 μCi) (71). The glands are washed and then incubated in 2 ml of Krebs-Ringer buffer containing 2 mg glucose and test agents. The glands are homogenized in 0.1 M HCl and heated for 15 min at 100°C; tris buffer is added and the mixture is neutralized. Cyclic AMP is isolated by $BaSO_4$ precipitation. Recoveries can be determined as previously described. Glands from control incubations (30 min) contain about 35 cpm of $[^3H]$-cyclic AMP, while after incubations with hypothalamic extract or prostaglandin E_1 these levels are increased to 10-fold and 7-fold, respectively. Endogenous cyclic AMP is increased in the presence of these agents to 10-fold and 30-fold, respectively, suggestive of compartmentalization of labeled adenine nucleotides, at least with respect to stimulation by prostaglandin E_1 [71].

7. Other Tissues

Kuehl et al. [53] investigated the accumulation of $[^{14}C]$-cyclic AMP in various tissues using the methodology described for testes prelabeled with adenine [69] and similarly labeled ovaries [70]. Prostaglandin E_1 causes an increase to more than 3-fold in levels of $[^{14}C]$-cyclic AMP in rat thymus, rat cerebral cortex, rat lung, rat seminal vesicle, bovine corpus luteum, and uterus from pregnant mouse.

Stimulations of 2-fold or less were reported with prostaglandin E_1 and rat prostrate gland, kidney, testes, and heart labeled with $[^{14}C]$-adenine.

Vasopressin is reported to cause the accumulation of $[^{14}C]$-cyclic AMP in adenine-prelabeled kidney slices [11]. Insulin-releasing agents elevate the level of $[^{14}C]$-cyclic AMP derived from labeling with radioactive adenine and the level of endogenous cyclic AMP in pancreatic islets to a similar extent [72]. Adenine nucleotides of prostrate gland can be labeled by incubation with radioactive adenosine [73]. Subsequent incubations with theophylline and testerone did not increase the concentration of radioactive cyclic AMP.

C. Perfused Organs

Investigation of the accumulation of cyclic AMP in perfused organs and *in vivo* by prelabeling techniques has been limited essentially to heart. Such techniques might well provide valuable information on the role of adenine- and adenosine-uptake processes in adenine nucleotide metabolism in whole animals.

1. Heart

Perfused hearts incorporate radioactive adenine,

hypoxanthine, inosine, and adenosine into intracellular adenine nucleotides [74-76]. Incorporation of adenosine is 15 to 20-fold that of inosine, and results in selective labeling of ATP and ADP [76]. The specific activity of the cardiac 5'-AMP is about half that of ATP-ADP.

The prelabeling technique with $[^{14}C]$-adenosine has been applied to the study of accumulation of cyclic AMP in perfused frog ventricles [79] as follows.

Frog ventricles are perfused for 90 min with a solution containing 0.03 μmoles of $[^{14}C]$-adenosine (2 μCi), and then thoroughly washed. Most of the radioactivity in the slice is associated with ATP and ADP. After conclusion of the experiments, tissue is frozen and high specific activity $[^{3}H]$-cyclic AMP is added to monitor recovery, followed by homogenization with TCA and purification of cyclic AMP, first on an anion-exchange column, and finally by thin-layer chromatography. The cyclic AMP fraction thus purified is analyzed for $[^{14}C]$-cyclic AMP, endogenous cyclic AMP and recovery of $[^{3}H]$-cyclic AMP by high-pressure liquid chromatography [77] (see also Chapter 3). The specific activities of cyclic AMP and ATP in the frog ventricles are of the same order of magnitude.

Radioactive cyclic AMP has also been determined

in adenosine-labeled guinea pig heart [78] as follows.

Hearts are back-perfused with 30-35 ml of recirculating Krebs-Henseleit solution fortified with amino acids, containing 1 μmolar [^{3}H]-adenosine (1 mCi) for 30 min, washed thoroughly, then perfused under control and overload conditions for 10 min. The frozen left-ventricles are homogenized with perchloric acid containing carrier cyclic AMP and [^{14}C]-cyclic AMP. Cyclic AMP is isolated by the method of Krishna [23].

2. Other Organs

Perfused liver rapidly incorporates radioactive adenine or hypoxanthine into intracellular adenine nucleotides [27]. After perfusion with Ringer lactate solution containing 5 μM [^{14}C]-adenine (53 Ci/mole) for 5 min at 24°C and further perfusion at 0°C for 8 min to remove any unincorporated adenine, the liver has incorporated about 80% of the perfused radioactive adenine into adenine nucleotides. In contrast to studies with other tissues, 75% of this radioactivity is associated with 5'-AMP, and only 25% with ATP and ADP. Translocation of radioactive purine bases from liver to erythrocytes, and the reverse, are demonstrated [27].

Injection of radioactive adenine into the carotid artery, or directly into the cisternae of rat brain, results in labeling of brain adenine nucleotides [2]. In contrast to results obtained *in vitro* with slices, a large portion (40%) of the radioactive adenine is not present in the free adenine mononucleotides.

The labeling of adenine nucleotides in perfused adrenal glands with [^{3}H]-adenosine, and the subsequent presence of radioactive cyclic AMP in the perfusate, has been reported in an abstract [79]. The labeled nucleotides were present in isolated chromaffin granules of the adrenal medulla.

Mouse ascites tumor cells contain labeled nucleotides after intraperitoneal administration of radioactive adenine [80].

III. CONCLUSIONS

The prelabeling technique for the study of adenine nucleotides, in particular cyclic AMP, has been used with various biological preparations. Selective labeling of intracellular compartments of adenine nucleotides and of specific cell types in tissue slices and whole organs will probably be established as a common, if not ubiquitous, phenomenon. Such compartmentalization of adenine nucleotides can be studied most profitably by the prelabeling technique, once the parameters for selective labeling of such compartments are ascertained. Short labeling times with low concentrations of adenine or adenosine may prove to "selectively" label the pools of adenine nucleotides that serve as precursors for hormone-elicit accumulations of cyclic AMP. The physiological significance of these pools, and their associated enzymes, adenylate cyclase, adenine phosphoribosyltransferase, and adenosine kinase is unknown.

NOTE ADDED IN PROOF

The accumulation of radioactive cyclic AMP in intact cells labeled by a prior incubation with radioactive adenine has recently been reported in fat cells with phosphodiesterase inhibitors [81], and as affected by sodium and potassium ions [82], in adrenal cells as affected by calcium ions [83], in rabbit brain slices as affected by histamine and lithium ions [84], and in corpus luteum [85]. The efflux of radioactive compounds from [^{14}C]-adenine-labeled slices of guinea pig cerebral cortex during electrical stimulation has been studied [86].

ACKNOWLEDGMENT

I thank Drs. B. Hamprecht and J. Schultz for their assistance with the section of glioma cells and Drs. Schultz and M. Huang for their assistance with certain portions of the section on brain slices.

REFERENCES

[1] H. Shimizu and J. Daly, in Methods in Neurochemistry, F. Rainer, ed. Marcel Dekker, New York, in press.

[2] H. Shimizu, S. Tanaka, and T. Kodama, J. Neurochem., in press.

[3] M. T. Hakala, L. N. Kenny, and H. K. Slocum, Federation Proc., 30, 679Abs (1971).

[4] J. Hochstadt-Ozer, Federation Proc., 30, 1062Abs (1971).

[5] H. Holmsen, H. J. Day, and M. A. Pimentel, Biochim. Biophys. Acta, 186, 244 (1969).

[6] J. F. Kuo and I. K. Dill, Biochem. Biophys. Res. Commun., 32, 333 (1968).

[7] J. F. Kuo, Biochem. Pharmacol., 18, 757 (1969).

[8] J. F. Kuo, Biochim. Biophys. Acta, 208, 509 (1970).

[9] J. F. Kuo and E. C. DeRenzo, J. Biol. Chem., 244, 2252 (1969).

[10] G. Krishna, S. Hynie, and B. B. Brodie, Proc. Natl. Acad. Sci. U.S., 58, 884 (1968).

[11] J. L. Humes, M. Rounbehler, and F. A. Kuehl, Anal. Biochem., 32, 210 (1969).

[12] J. Forn, P. S. Schonhofer, I. F. Skidmore, and G. Krishna, Biochim. Biophys. Acta, 208, 304 (1970).

[13] P. S. Schonhofer and I. F. Skidmore, Pharmacology 6, 109 (1971).

[14] C. H. Hollenberg and R. L. Patten, Metabolism, 19, 856 (1970).

[15] A. M. Williams and G. A. LePage, Cancer Res., 18, 554 (1958).

[16] J. Schultz, B. Hamprecht, and J. Daly, Proc. Natl. Acad. Sci. U.S., in preparation.

[17] A. G. Gilman, Proc, Natl. Acad. Sci. U.S., 67, 305 (1970).

[18] H. Holmsen and M. C. Rozenberg, Biochim. Biophys. Acta, 155, 326 (1968).

[19] H. Holmsen, H. J. Day, and E. Storm, Biochim. Biophys. Acta, 186, 254 (1969).

[20] M. C. Rozenberg and H. Holmsen, Biochim. Biophys. Acta, 155, 342 (1968).

[21] H. Holmsen and M. C. Rozenberg, Biochim. Biophys. Acta, 157, 266 (1968).

[22] R. L. Bigdahl, N. R. Marquis, and P. A. Tavormina, Biochem. Biophys. Res. Commun., 37, 409 (1969).

[23] G. Krishna, B. Weiss, and B. B. Brodie, J. Pharmacol. Exptl. Therap., 163, 379 (1968).

[24] R. J. Haslam and A. Taylor, Biochem. J., 125, 377 (1971).

[25] R. J. Haslam and A. Taylor, in Platelet Aggregation, J. Caen, ed., Masson et Cie, Paris, 1971, p. 85.

[26] J. Moskowitz, J. P. Harwood, W. D. Reid, and G. Krishna, Federation Proc., 29, 602Abs (1970); J. P. Harwood, J. Moskowitz, and G. Krishna, Federation Proc., 30, 285Abs (1971).

[27] J. B. Pritchard, F. Chavez-Peon, and R. D. Berlin, Am. J. Physiol., 219, 1263 (1970).

[28] O. M. Rosen and J. Erlichman, Arch. Biochem. Biophys., 133, 171 (1969).

[29] H. R. Bourne, K. L. Melmon, and L. M. Lichtenstein, Science, 173, 743 (1971).

[30] H. R. Bourne, R. I. Lehrer, M. J. Cline, and K. L. Melmon, J. Clin. Invest., 50, 920 (1971).

[31] K. A. C. Elliot, Tissue Slice Technique, in Methods in Enzymology, Vol. 1, (S. P. Colowick and N. O. Kaplan, eds.) Academic Press, New York, 1955, p. 3.

[32] H. McIlwain and R. Rodknight, Practical Neurochemistry, Churchill, London, 1962.

[33] J. N. Santos, K. W. Hempstead, L. E. Kopp, and R. P. Miech, J. Neurochem., 15, 367 (1968).

[34] S. K. Sharma and U. N. Singh, J. Neurochem., 17, 305 (1970).

[35] I. Held and W. Wells, J. Neurochem., 16, 529 (1969).

[36] H. Shimizu, J. W. Daly, and C. R. Creveling, J. Neurochem., 16, 1609 (1969).

[37] H. Shimizu and J. Daly, Biochim. Biophys. Acta, 222, 465 (1970).

[38] J. Forn and G. Krishna, Pharmacology, 5, 193 (1971).

[39] H. Shimizu, C. R. Creveling, and J. W. Daly, Adv. Biochem. Psychopharmacol., 3, 135 (1970).

[40] G. Krishna, J. Forn, K. Voigt, M. Paul, and G. L. Gessa, Adv. Biochem. Psychopharmacol., 3, 155 (1970).

[41] H. Shimzu, C. R. Creveling, and J. Daly, Proc. Natl. Acad. Sci. U.S., 65, 1033 (1970).

[42] H. Shimizu, S. Tanaka, T. Suzuki, and Y. Matsukac J. Neurochem., 18, 1157 (1971).

[43] R. Fumagalli, V. Bernareggi, F. Berti, and M. Trabucchi, Life Sci., 10, Part 1, 1111 (1971).

[44] M. Huang and J. Daly, J. Med. Chem., 15, 458 (1972).

[45] M. Chasin, I. Rivkin, F. Mamrak, S. G. Samaniego, and S. M. Hess, J. Biol. Chem., 246, 3037 (1971).

[46] H. Shimizu, C. R. Creveling, and J. Daly, J. Neurochem., 17, 441 (1971).

[47] M. Huang, H. Shimizu, and J. W. Daly, J. Med. Chem., 15, 462 (1972).

[48] H. Shimizu and J. Daly, European J. Pharmacol., in press.

[49] H. Shimizu, C. R. Creveling, and J. W. Daly, Mol. Phamacol., 6, 184 (1971).

[50] T. Kodama, Y. Matsukado, T. Suzuki, S. Tanaka, and H. Shimizu, Biochim. Biophys. Acta, 252, 165 (1971).

[51] M. Huang, H. Shimizu, and J. Daly, Mol. Pharmacol., 7, 155 (1971).

[52] I. Pull and H. McIlwain, Biochem. J., in press.

[53] F. A. Kuehl, Jr., J. L. Humes, V. J. Cirillo, and E. A. Ham, Advances in Cyclic Nucleotide Research, Vol I, (P. Greengard and R. Paoletti), eds., Raven Press, New York, in press.

[54] M. Huang and J. Daly, in preparation.

[55] S. Kakiuchi and T. W. Rall, Mol. Pharmacol., 4, 367 (1968).

[56] S. Kakiuchi and T. W. Rall, Mol. Pharmacol., 4, 379 (1968).

[57] H. Shimizu, M. Huang, and J. Daly, unpublished results.

[58] S. Kakiuchi, T. W. Rall, and H. McIlwain, J. Neurochem., 16, 485 (1969).

[59] J. Schultz and J. Daly, J. Biol. Chem., in preparation; see also J. Schultz and J. Daly, Abstr. V. Internatl. Congr. Pharmacol., San Francisco, 1972.

[60] H. Shimizu and H. Okayama, J. Biol. Chem., submitted.

[61] M. Huang and J. Daly, unpublished results.

[62] T. W. Rall and A. Sattin, Adv. Biochem. Psychopharmacol., 3, 113 (1970).

[63] A. Sattin and T. W. Rall, Mol. Pharmacol., 6, 13 (1970).

[64] T. W. Rall, Ann. N. Y. Acad. Sci., 185, 520 (1971).

[65] F. Bloom, M. Huang, and J. Daly, unpublished results [cited in J. Daly, M. Huang, and H. Shimizu, Advances in Cyclic Nucleotide Research, Vol. I (P. Greengard, G. A. Robison, and R. Paoletti, eds.), Raven Press, New York, in press].

[66] A. G. Gilman and M. Nirenberg, Proc. Natl. Acad. Sci. U.S., 68, 2165 (1971).

[67] R. G. Clark and J. P. Perkins, Proc. Natl. Acad. Sci. U.S., 68, 2757 (1971).

[68] P. J. LaRaia and W. J. Reddy, Biochim. Biophys. Acta, 177, 189 (1969).

[69] F. A. Kuehl, Jr., D. J. Patanelli, J. Tarnoff, and J. L. Humes, Biol. Reprod., 2, 154 (1970).

[70] F. A. Kuehl, Jr., J. L. Humes, J. Tarnoff, V. J. Cirillo, and E. A. Ham, Science, 169, 883 (1970)

[71] U. Zor, T. Kaneko, H. P. G. Schneider, S. M. McCann, and J. B. Field, J. Biol. Chem., 245, 2883 (1971).

[72] J. F. Kuo, personal communication.

[73] C. G. Smith, J. A. Thomas, and M. G. Mawhinney, Pharmacologist, 13, 287 abs. (1971).

[74] K. K. Tsuboi and N. M. Buckley, Circulation Res., 16, 343 (1965).

[75] M. I. Jacob and R. M. Berne, Am. J. Physiol., 198, 322 (1960).

[76] M. S. Liu and H. Feinberg, Am. J. Physiol., 220, 1242 (1971).

[77] G. Brooker, J. Biol. Chem., in press.

[78] S. S. Schreiber, I. L. Klein, M. Oratz, and M. A. Rothschild, J. Mol. Cell. Cardiology, 2, 55 (1971).

[79] P. Stevens and K. VanDyke, Pharmacologist, 13, 287 abs. (1971).

[80] A. M. Williams and G. A. LePage, Cancer Res., 18, 562 (1958).

[81] L. R. Mandel, Biochem. Pharm., 20, 3413 (1971).

[82] J. Gorski and G. Krishna, Federation Proc., 31, 861 abs (1972).

[83] G. Sayers, R. J. Beall and S. Seelig, Science, 175, 1131 (1972).

[84] J. Forn and F. G. Valdecasas, Biochem. Pharm., 20, 2773 (1971).

[85] S. Goldstein and J. M. Marsh, Federation Proc., 31, 861 abs (1972).

[86] I. Pull and H. McIlwain, Biochem. J., 126, 965 (1972).

AUTHOR INDEX

Numbers in brackets are reference numbers and indicate that an author's work is referred to although his name is not cited in the text. Underlined numbers give the page on which the complete reference is listed.

W

Z

SUBJECT INDEX

T

U

V

Z